Collins

INTERNATIONAL PRIMARY SCIENCE

Progress Book 3

Student's Book

William Collins' dream of knowledge for all began with the publication of his first book in 1819.
A self-educated mill worker, he not only enriched millions of lives, but also founded a flourishing publishing house.
Today, staying true to this spirit, Collins books are packed with inspiration, innovation and practical expertise.
They place you at the centre of a world of possibility and give you exactly what you need to explore it.

Collins. Freedom to teach.

Published by Collins

An imprint of HarperCollins*Publishers*
The News Building, 1 London Bridge Street, London, SE1 9GF, UK

HarperCollins*Publishers*
Macken House, 39/40 Mayor Street Upper, Dublin 1, D01 C9W8, Ireland

Browse the complete Collins catalogue at
www.collins.co.uk

10 9 8 7 6 5 4 3 2 1

ISBN 978-0-00-865487-0

British Library Cataloguing-in-Publication Data
A catalogue record for this publication is available from the British Library.

Author: Tracy Wiles
Publisher: Elaine Higgleton
Product manager: Holly Woolnough
Project manager: Just Content Ltd
Copy editor: Karen Williams
Answer checker: Tanya Solomons
Proofreader: Catherine Dakin
Cover designer: Gordon MacGilp
Cover illustration: Ann Paganuzzi
Typesetter: David Jimenez
Illustrator: Ann Paganuzzi
Production controller: Lyndsey Rogers
Printed and bound in Great Britain by Martins the Printers

With thanks to the following teachers for reviewing materials in proof and providing valuable feedback: Sylvie Meurein, Nilai International School; Gabriel Kehinde, Avi-Cenna International School; and with thanks to the following teachers who provided feedback during the early development stage: Najihah binti Roslan, Nilai International School.

FSC
www.fsc.org
MIX
Paper | Supporting responsible forestry
FSC™ C007454

This book is produced from independently certified FSC™ paper to ensure responsible forest management.
For more information visit: harpercollins.co.uk/green

Contents

Topic 3 Humans and other animals

Topic 4 Materials

Topic 5 States of matter

Topic 6 Forces, light and magnets

Topic 7 Earth, Moon and Sun

How to use this book

This book is full of questions. Each set of questions can be completed at the end of each unit.

The questions allow you to practise the things you've learned. They will help you understand topics that you might need more practice of. They will also show you the topics that you are most confident with. Your teacher can use your answers to give you feedback and support.

At the end of each test, there is a space to put the date that you completed it. There is also a blank box. Your teacher might use it to:

- sign, when they have marked your answers
- write a short comment on your work.

Date: ______________________	Now look at and think about each of the *I can* statements.	

Pages 7 to 15 include a list of *I can* statements. Once you have finished each set of questions, turn to the *I can* statements. Think about each statement: how easy or hard did you find the topic? For each statement, colour in the face that is closest to how you feel:

☺ I can do this ☹ I'm getting there ☹ I need some help.

There are three longer Summative Assessments in the book. These can be completed after each block of topics.

I can statements

At the end of each unit, think about each of the *I can* statements and how easy or hard you find the topic. For each statement, colour in the face that is closest to how you feel.

TOPIC 1 Living and non-living things	
1.1 Scientific enquiry	**Date:**
I can use the five main types of scientific enquiry: research, fair testing, observing over time, identifying and classifying, and pattern seeking.	☺ 😐 ☹
1.2 Life processes	**Date:**
I know that plants and animals are living things.	☺ 😐 ☹
I know that movement, growth, nutrition and reproduction are life processes.	☺ 😐 ☹
1.3 Is it living?	**Date:**
I know that some things are living and others are non-living.	☺ 😐 ☹
I know that plants and animals are living things.	☺ 😐 ☹
I know that rocks and soil are non-living things.	☺ 😐 ☹
1.4 Fossils	**Date:**
I know that fossils are the dead remains of plants and animals.	☺ 😐 ☹
I know that fossils are rock impressions of skeletons.	☺ 😐 ☹
I know that imprints show the outer shape of plants and animals, and that they are less common fossils.	☺ 😐 ☹
1.5 Food chains	**Date:**
I know that food chains show the feeding relationships in a habitat.	☺ 😐 ☹

I know that all food chains start with a producer.	☺	😐	☹
I know that animals are consumers because they eat plants and other animals.	☺	😐	☹
TOPIC 2 Plants			
2.1 Parts of plants	**Date:**		
I know that plants have different parts.	☺	😐	☹
I know that each part of a plant has a different role.	☺	😐	☹
I can identify the roots, stem, leaves and flowers on a plant.	☺	😐	☹
2.2 Plants need water	**Date:**		
I know that plants need water to grow.	☺	😐	☹
I know that water is taken in through the roots and transported through the stem to other parts of the plant.	☺	😐	☹
2.3 Plants need sunlight	**Date:**		
I know that humans and other animals need plants for food.	☺	😐	☹
I know that plants need sunlight to make food and to grow.	☺	😐	☹
2.4 Plants need warmth	**Date:**		
I know that plant growth is affected by temperature.	☺	😐	☹
I know that plants need warmth to grow.	☺	😐	☹
2.5 Healthy plants	**Date:**		
I know that plants need healthy roots, stems and leaves to grow well.	☺	😐	☹
I know that soil can be improved by using fertilisers.	☺	😐	☹
I know that plants can be irrigated in dry areas.	☺	😐	☹

2.6 Bees need flowering plants	Date:
I know that bees drink nectar from flowers.	☺ 😐 ☹
I know that bees transport pollen from flower to flower, which helps the flowers to produce seeds.	☺ 😐 ☹
I know that bees all across the world are dying out and that we need to do something to save them.	☺ 😐 ☹
TOPIC 3 Humans and other animals	
3.1 Models and representations	**Date:**
I know that two-dimensional and three-dimensional models help us to understand science better.	☺ 😐 ☹
3.2 Classifying animals	**Date:**
I know that scientists classify animals into groups.	☺ 😐 ☹
I know that each group of animals shares common features.	☺ 😐 ☹
I can identify the six main groups of animals (mammals, reptiles, insects, birds, amphibians, fish).	☺ 😐 ☹
3.3 Growing and changing	**Date:**
I know that humans and other animals grow and change.	☺ 😐 ☹
I know that some young animals look very different when they grow into adults.	☺ 😐 ☹
3.4 Your brain	**Date:**
I know the brain controls the body systems, responds to our senses, and allows us to think, remember and feel emotions.	☺ 😐 ☹
I know the brain has three main parts.	☺ 😐 ☹
I know the brain is very fragile and needs to be protected.	☺ 😐 ☹

3.5 Your heart	Date:
I know that the heart is a pump made of muscle.	☺ 😐 ☹
I know that the heart pumps blood around our body.	☺ 😐 ☹
I know that the heart pumps faster when we exercise.	☺ 😐 ☹
3.6 Your stomach and intestines	**Date:**
I know that the digestive system starts in the mouth with chewing food.	☺ 😐 ☹
I know that food then gets broken down into smaller pieces in the stomach and small intestines.	☺ 😐 ☹
I know that food enters the blood in the small intestines.	☺ 😐 ☹
I know that unwanted food leaves the body through the large intestine.	☺ 😐 ☹
3.7 Your lungs	**Date:**
I know that lungs take oxygen in to the body and release carbon dioxide out of the body.	☺ 😐 ☹
I know that my lungs are protected by my ribcage.	☺ 😐 ☹
3.8 Surgery and science	**Date:**
I know that keyhole surgery means a quicker recovery time, a smaller chance of infection, less blood loss and less pain after the surgery.	☺ 😐 ☹
I know that CT and MRI scans allow a surgeon to see inside a person's body without having to operate on them.	☺ 😐 ☹
I know that surgeons use ultrasound scanners during surgery.	☺ 😐 ☹

TOPIC 4 Materials	
4.1 Types of materials	**Date:**
I know that materials are all around us.	☺ 😐 ☹
I know that materials can be solids, liquids or gases.	☺ 😐 ☹
4.2 Materials from Earth	**Date:**
I know that all materials we use come from the Earth.	☺ 😐 ☹
I know that oil, natural gas and pure metals are materials that come from rocks.	☺ 😐 ☹
TOPIC 5 States of matter	
5.1 Solids and liquids	**Date:**
I know that it is important to use standard measurements to make investigations reliable and fair.	☺ 😐 ☹
I know that it is important to take safety precautions when conducting an experiment.	☺ 😐 ☹
5.2 Solid or liquid?	**Date:**
I know that solid materials have a definite shape.	☺ 😐 ☹
I know that some solid materials are brittle and break easily.	☺ 😐 ☹
I know that some solids are malleable and can be squashed into different shapes without breaking.	☺ 😐 ☹
I know that liquids take the shape of the container they are in.	☺ 😐 ☹
I know that liquids flow easily and cannot easily be compressed.	☺ 😐 ☹
I know that some liquids pour quickly while others pour slowly.	☺ 😐 ☹
5.3 Mixtures	**Date:**
I know that a mixture is made of two or more materials mixed together.	☺ 😐 ☹

I know that the materials in a mixture retain their own properties.	
I know that the materials in a mixture can easily be separated.	
I know that some methods of separating mixtures are sieving, filtering and using a magnet.	
5.4 Mixing and separating solids	**Date:**
I know that individual materials in a mixture are not joined together, which makes it easy to separate them.	
I know that a sieve can be used to separate solids of different sizes.	
I know that a magnet can be used to separate magnetic materials from other materials.	
5.5 Solids in liquids	**Date:**
I know that a colander can be used to remove large solids from water.	
I know that filtration is the method used to remove small pieces of solid from a liquid.	
5.6 Dissolving solids in liquids	**Date:**
I know that some materials dissolve when you mix them with water.	
I know that you cannot see dissolved materials in water.	
I know that if a material has dissolved in a liquid, the liquid will always be clear. It may or may not be coloured.	
TOPIC 6 Force, light and magnets	
6.1 Friction investigation	**Date:**
I know that bar charts are useful when comparing data in different categories.	
6.2 Gravity – a natural force	**Date:**
I know that gravity keeps us on the ground and makes things fall back down to Earth.	

I know that gravity is a natural force which pulls everything towards the centre of the Earth in a downwards direction.	☺ 😐 ☹
6.3 Measuring forces	**Date:**
I know that gravity is a downward force that acts on all objects.	☺ 😐 ☹
I know that we can measure gravity using a force meter.	☺ 😐 ☹
I know that Newtons are a measurement of force.	☺ 😐 ☹
6.4 Friction	**Date:**
I know that friction is a force that slows things down.	☺ 😐 ☹
I know that friction acts in the opposite direction to the movement.	☺ 😐 ☹
I know that friction can be useful.	☺ 😐 ☹
6.5 Will it move?	**Date:**
I know that friction occurs when objects are in contact with each other.	☺ 😐 ☹
I know that even surfaces that look smooth have tiny grooves and ridges on them.	☺ 😐 ☹
I know that there is more friction on a surface which is rough.	☺ 😐 ☹
I know that friction is a force that slows things down.	☺ 😐 ☹
6.6 Letting light through	**Date:**
I know that transparent materials let light through.	☺ 😐 ☹
I know that translucent materials let some light through.	☺ 😐 ☹
I know that opaque materials do not let any light through.	☺ 😐 ☹

6.7 Shadows	Date:
I know that a shadow is a dark area that forms when the path of light is blocked.	☺ 😐 ☹
I know that the size of a shadow changes when the position of the light source changes.	☺ 😐 ☹
6.8 Magnets	**Date:**
I know that the two ends of a magnet are known as poles (north and south poles).	☺ 😐 ☹
I know that the opposite poles of a magnet attract each other.	☺ 😐 ☹
I know that the same poles repel each other.	☺ 😐 ☹
6.9 Magnetic and non-magnetic	**Date:**
I know that not all materials are magnetic.	☺ 😐 ☹
I know that magnets attract some types of metals.	☺ 😐 ☹
I know that materials which are not magnetic are called non-magnetic materials.	☺ 😐 ☹
6.10 The compass	**Date:**
I know that a compass needle is a magnet that is attracted to magnetic north.	☺ 😐 ☹
I know that a compass needle always points towards magnetic north.	☺ 😐 ☹
I know that true north is 500 kilometres north of magnetic north.	☺ 😐 ☹
TOPIC 7 Earth, Moon and Sun	
7.1 Moon changes	**Date:**
I know that the Moon reflects light from the Sun.	☺ 😐 ☹
I know that we see different shapes or phases of the Moon at different times of the lunar month.	☺ 😐 ☹
I know that the Moon rises in the east and sets in the west.	☺ 😐 ☹

7.2 The moving Earth and Moon	Date:
I know that the Moon orbits the Earth once roughly every 29½ days.	☺ 😐 ☹
I know that the Earth orbits the Sun every 365¼ days (one year).	☺ 😐 ☹
I know that the Earth rotates once every 24 hours.	☺ 😐 ☹
7.3 Shape of the Earth, Moon and Sun	**Date:**
I know that the Moon is shaped like a flattened sphere.	☺ 😐 ☹
I know that the Earth is almost spherical in shape, but it bulges at the equator and is flatter at the poles.	☺ 😐 ☹
I know that the Sun is almost perfectly spherical.	☺ 😐 ☹
7.4 Astronomy then and now	**Date:**
I know that astronomers are scientists who study the Universe and all of the objects within it.	☺ 😐 ☹
I know that scientists have proven that the Earth moves around the Sun.	☺ 😐 ☹
I know that modern scientists all over the world are still learning about the Solar System.	☺ 😐 ☹
I know that scientists live, work and carry out lots of experiments on the International Space Station (ISS).	☺ 😐 ☹

1 Investigate whether plants need water to grow or if they can grow in any type of liquid.

a What do you need to research?

____

b How can you make your investigation a fair test?

____

c Identify any possible risks before starting the investigation.

____

2 Look at the pictures of plants from the completed investigation. Record your observations in the chart.

Plant pot	Observation
Pot 1 (mouth wash)	
Pot 2 (milk)	
Pot 3 (orange juice)	
Pot 4 (fizzy juice)	
Pot 5 (water)	

3 What conclusions can you make from the results of the investigation? Can you identify any patterns?

____

____

4 Investigate whether seeds need water, warmth and light to grow.

a What do you need to research?

____

b How can you make your investigation a fair test?

____

c Identify any possible risks before starting the investigation.

____

5 Look at the pictures of seeds from the completed investigation. Record your observations in the chart.

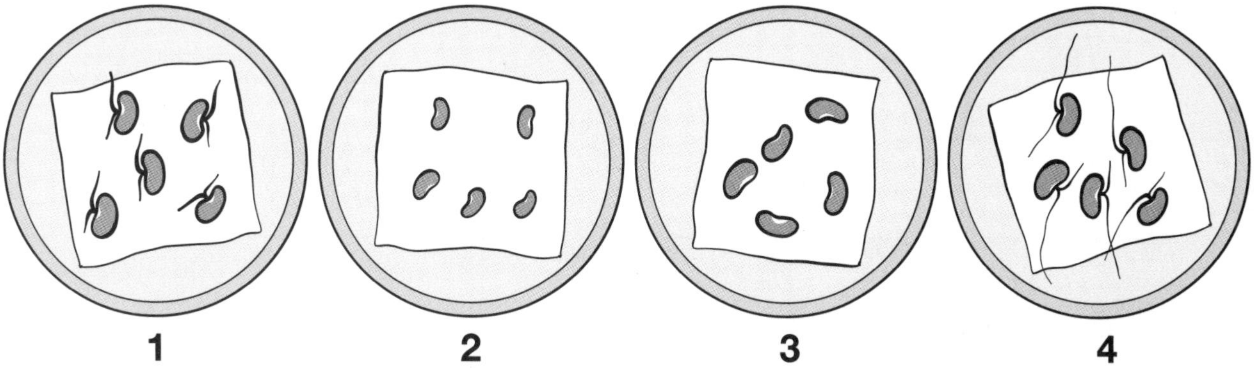

Seed dish	Observation
Dish 1 (control – has water, warmth and light)	
Dish 2 (no water)	
Dish 3 (no warmth)	
Dish 4 (no light)	

6 What conclusions can you make from the results of the investigation?

____

____

Date: ____________________

Now look at and think about each of the *I can* statements.

1 Complete the chart. Tick (✓) the box if the answer is 'yes'.

	Does it grow?	**Does it eat?**	**Does it move?**	**Does it reproduce?**	**Is it living?**

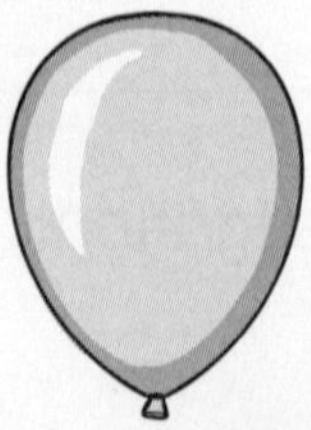					

2 Explain how you know that some of the things in question **1** are living.

____

____

____

____

____

____

3 Label the life process shown in each picture.

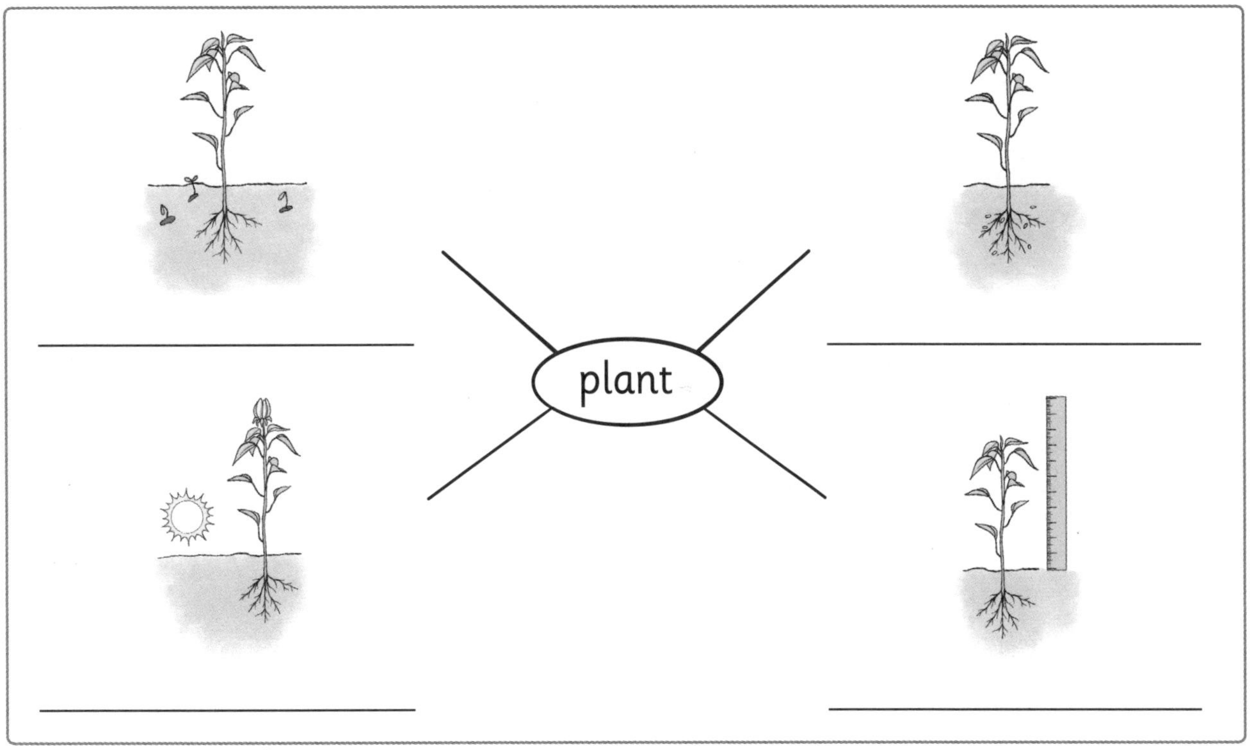

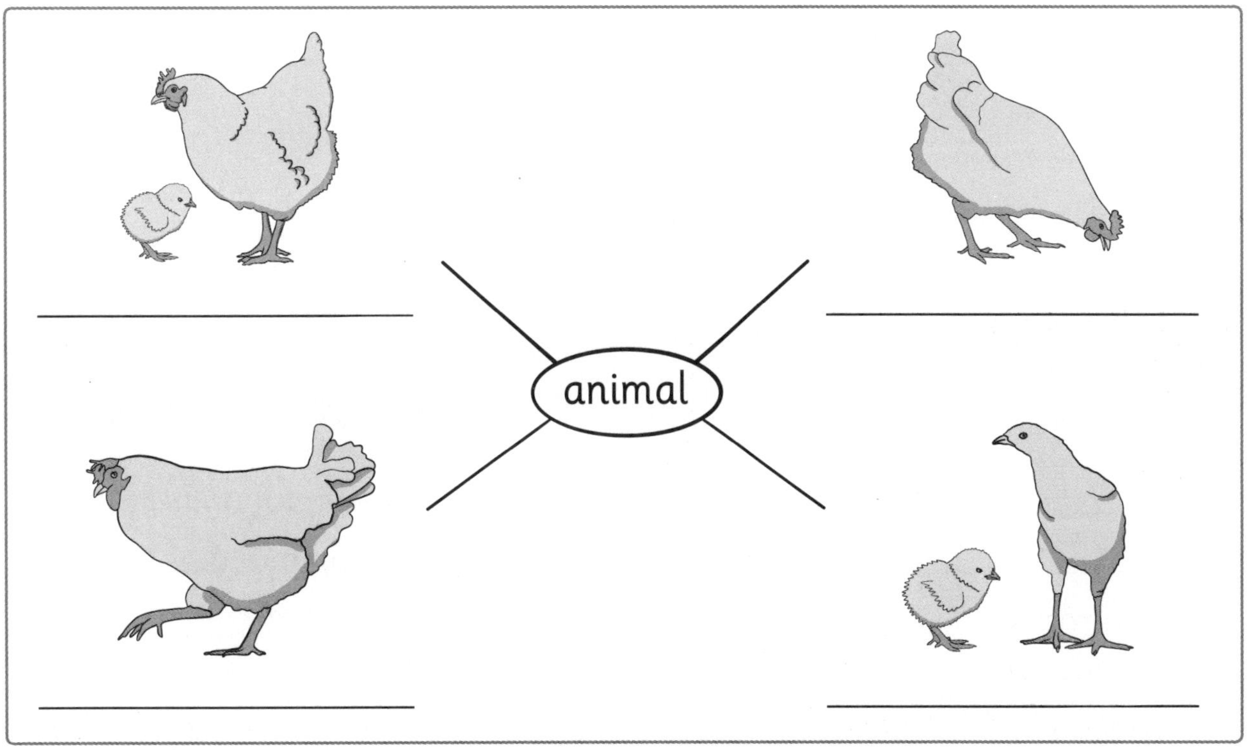

Now look at and think about each of the *I can* statements.

Date: ______________________

1 Write the words from the box in the correct columns.

tree spider rock grass book
bird car plastic bottle flower water

Living	Non-living

2 Which objects in the non-living column of question **1** are made from things that were once alive?

____

3 Which objects in the non-living column were never alive?

____

____

4 Use the words from the box to complete the sentences about how fossils are formed.

Fossils sediment imprints decays Palaeontologists pores

__________________ are scientists who dig fossils out of the rock.
__________________ are made when a plant or animal dies and is covered in __________________. The soft tissue __________________ quickly, but the bones take longer. The bones of a skeleton have spaces in them called __________________. The pores become full of mineral-rich water. Over millions of years, the minerals turn to crystals and then rock. The traces that are left are fossils. Some fossils are __________________ of animals and plants. These fossils show the outer shape of the animal or plant.

5 Look at the pictures. Identify the types of fossils. Tick (✓) the box if the answer is 'yes' to the questions.

Fossil	Is it an imprint or skeleton fossil?	Is the fossil of a plant or animal?
a 		
b 		
c 		
d 		
e 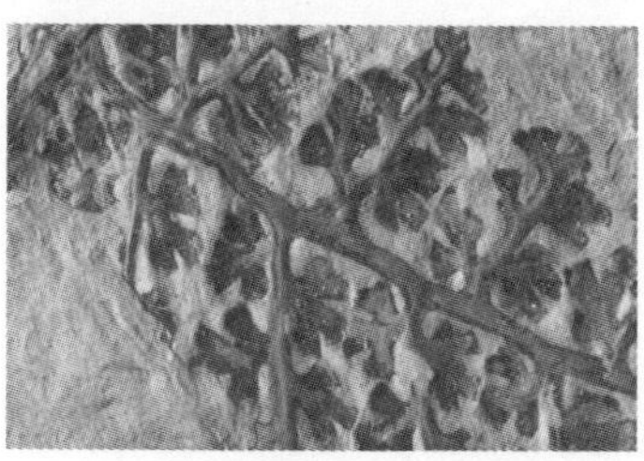		

Now look at and think about each of the *I can* statements.

Date: ______________________

1 Put the food chains in the correct order. Write the order under each food chain. Don't forget to add the arrows.

a

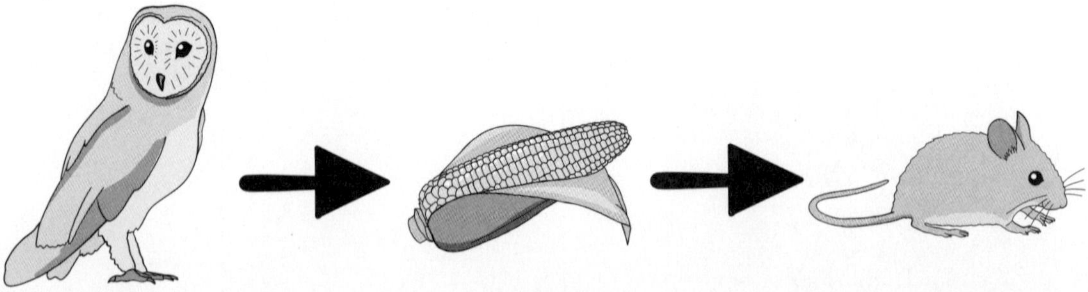

owl corn mouse

____

b

carrots fox rabbit

____

c

grasshopper snake frog grass

____

2 What type of living thing is always the producer in a food chain?

____

3 What type of living thing is always the consumer in a food chain?

____

4 Look at the food chains. Label the plants and animals by writing **producer** or **consumer**.

a

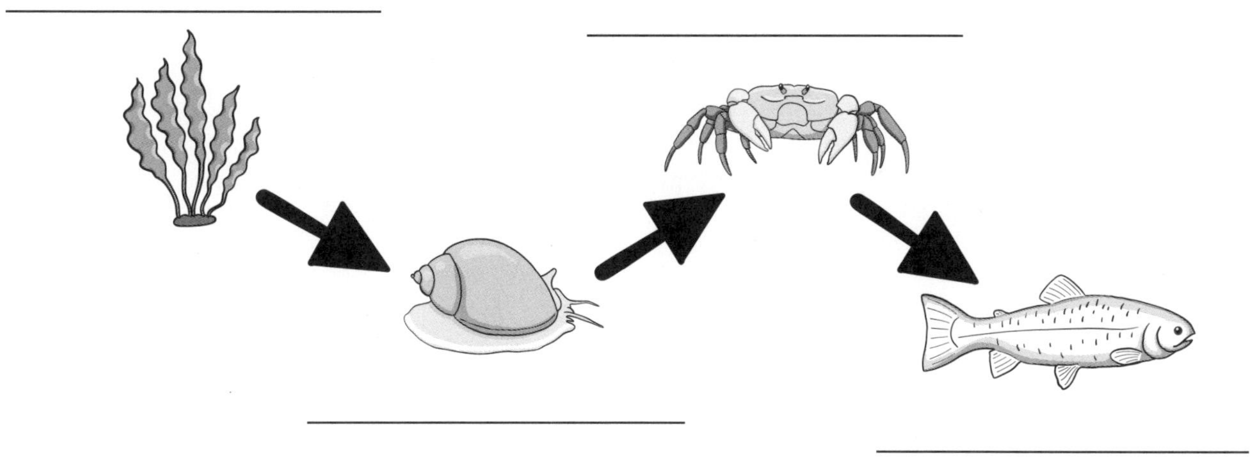

b

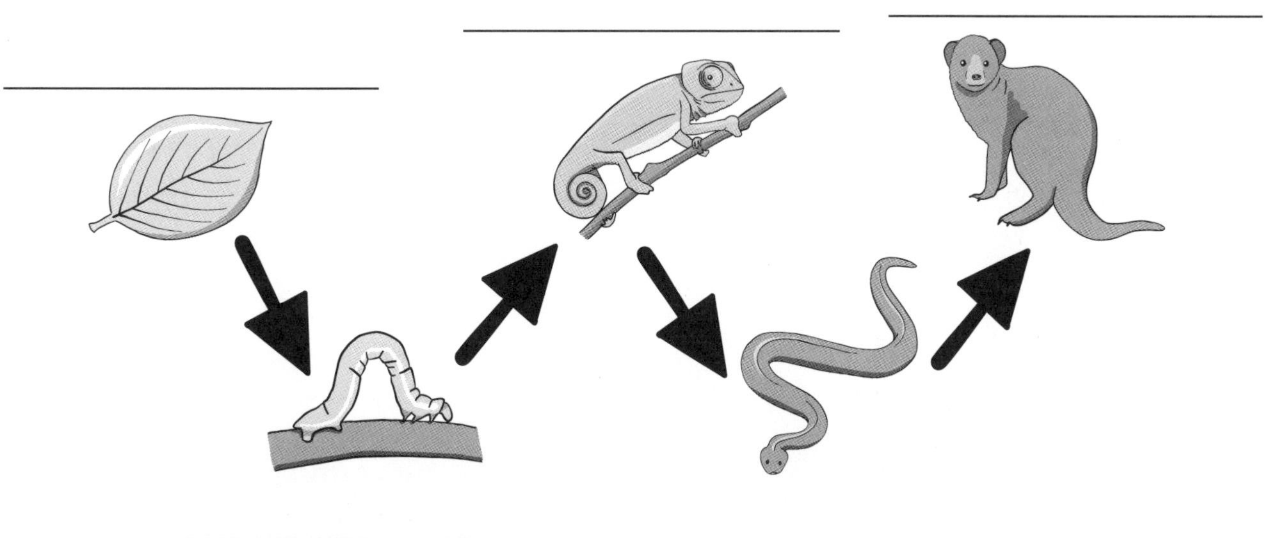

5 Circle the correct word to complete each sentence.

a Food chains show the **feeding** / **breathing** relationships in a habitat.

b Each organism in a food chain is important and helps the ecosystem to stay **unbalanced** / **balanced**.

c A lettuce plant is an example of a **producer** / **consumer**.

d A zebra is an example of a **producer** / **consumer**.

Now look at and think about each of the *I can* statements.

Date: ____________________

1 Draw a picture of a plant. Use the words in the box to label your plant.

stem roots flower leaf

Plant

2 Why would a plant put into a pot like this not grow?

____

____

____

____

____

____

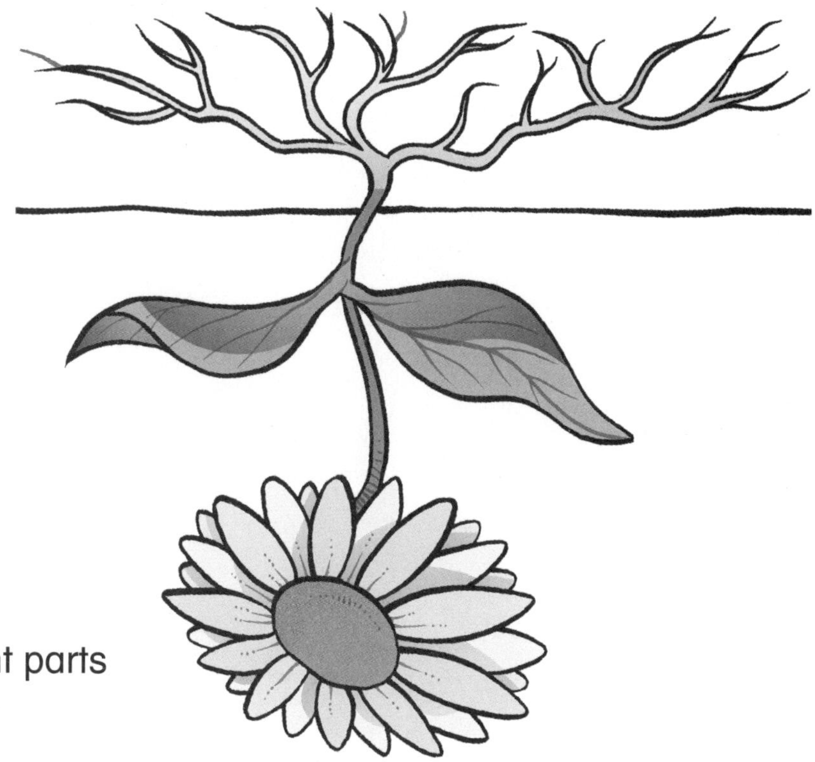

3 Explain why these different parts of a plant are important.

a roots

____

____

b stem

____

____

c leaves

____

____

d flowers

____

____

Date: ______________

Now look at and think about each of the *I can* statements.

1 What do plants need to grow? Circle the correct pictures.

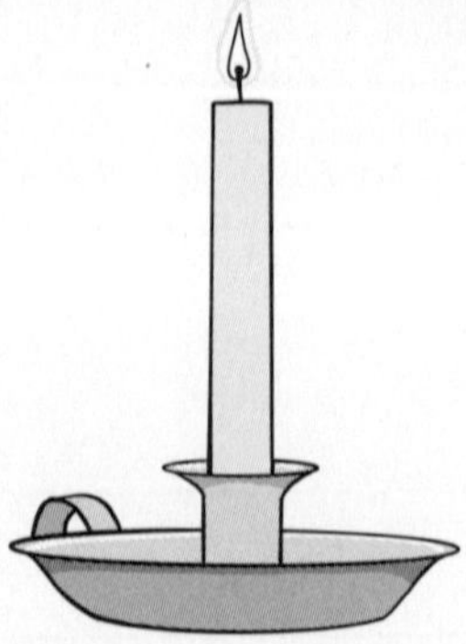

2 Circle the correct word to complete each sentence.

a Plants need **warmth** / **water** so that they do not dry out.

b Plants need **water** / **sunlight** so that they can make their own food.

c Plants need **sunlight** / **warmth** for seeds to grow.

3 Tick (✓) the sentence that explains what will happen to a plant if it does not get enough water.

The plant will stay firm and healthy. ☐

The plant will wilt and die. ☐

The plant will make its own water to live off. ☐

4 Look at the picture of a plant, then answer the questions.

a Circle the part of the plant that makes food.

b Draw arrows on the plant to show the way water is transported through the plant.

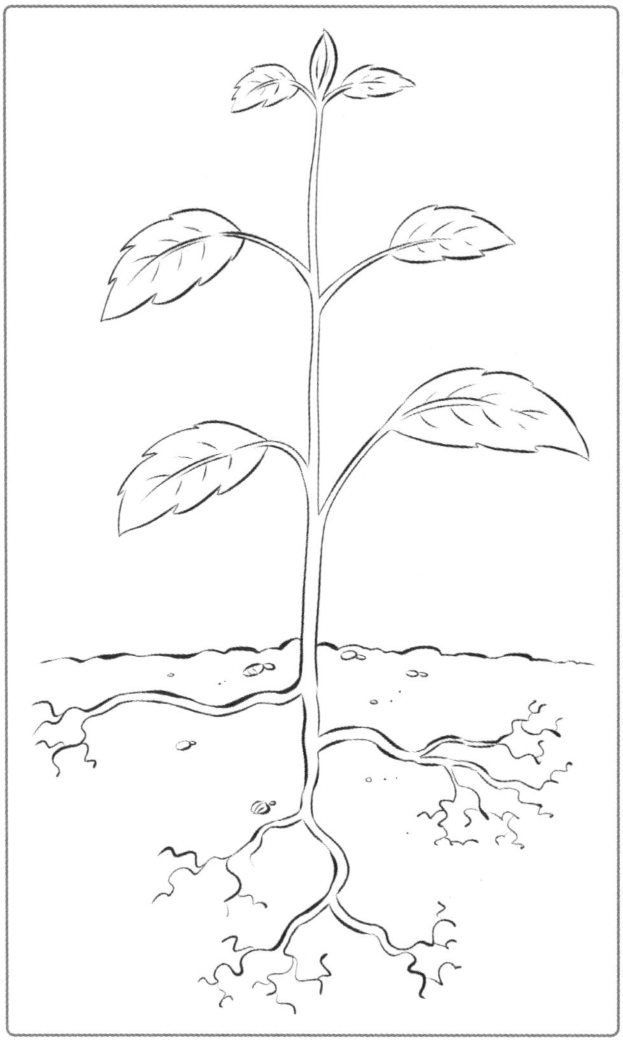

5 Look at the pictures of two plants below, then answer the questions.

Plant A **Plant B**

a Which plant is healthy? ____________________

b Which plant was placed in a dark place? ____________________

c Why is it important for plants to get sunlight?

____

d Why do humans and other animals need plants?

____

Date: ____________________

Now look at and think about each of the *I can* statements.

1 Write the words and phrases from the box in the correct columns.

poor soil   no light   no water   watered often
lots of sunlight   good soil   strong plant   weak plant
strong roots, stems and leaves
weak roots, stems and leaves

Healthy plant	Unhealthy plant

2 Use the information in the chart above to draw pictures of a healthy plant and an unhealthy plant.

Healthy plant	Unhealthy plant

3 Look at the pictures. Circle the fruit or vegetable in each pair that grew in soil that is rich in minerals.

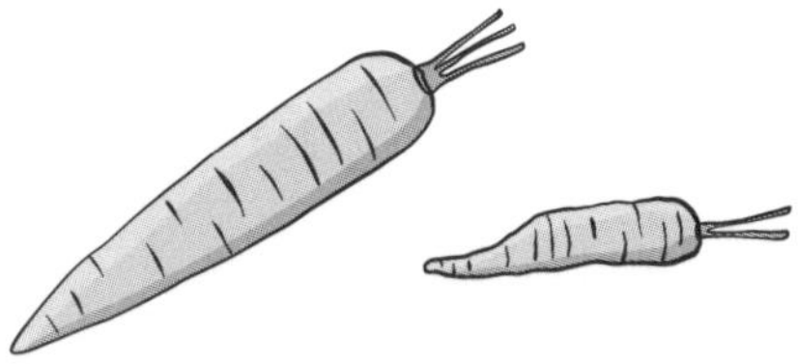

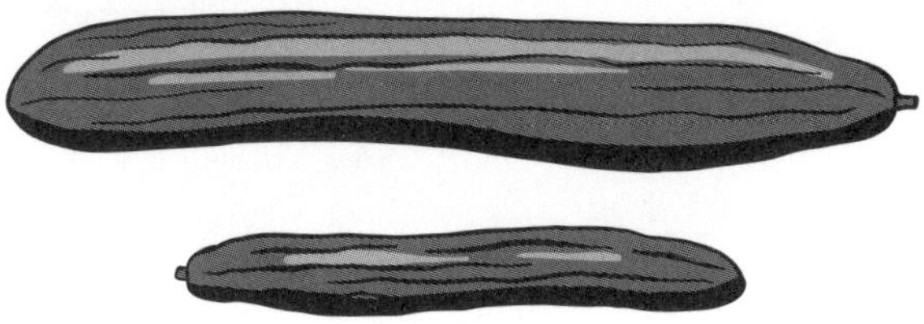

4 The photograph below shows irrigated fields in an area that used to be desert.

a Why do plants not usually grow well in desert areas? Circle the correct answers.

poor soil   good soil   poor climate   good climate

not enough water   plenty of water

b How has technology helped people to grow plants in desert areas?

____

Date: ____________________

Now look at and think about each of the *I can* statements.

1 Look at the picture, then answer the questions.

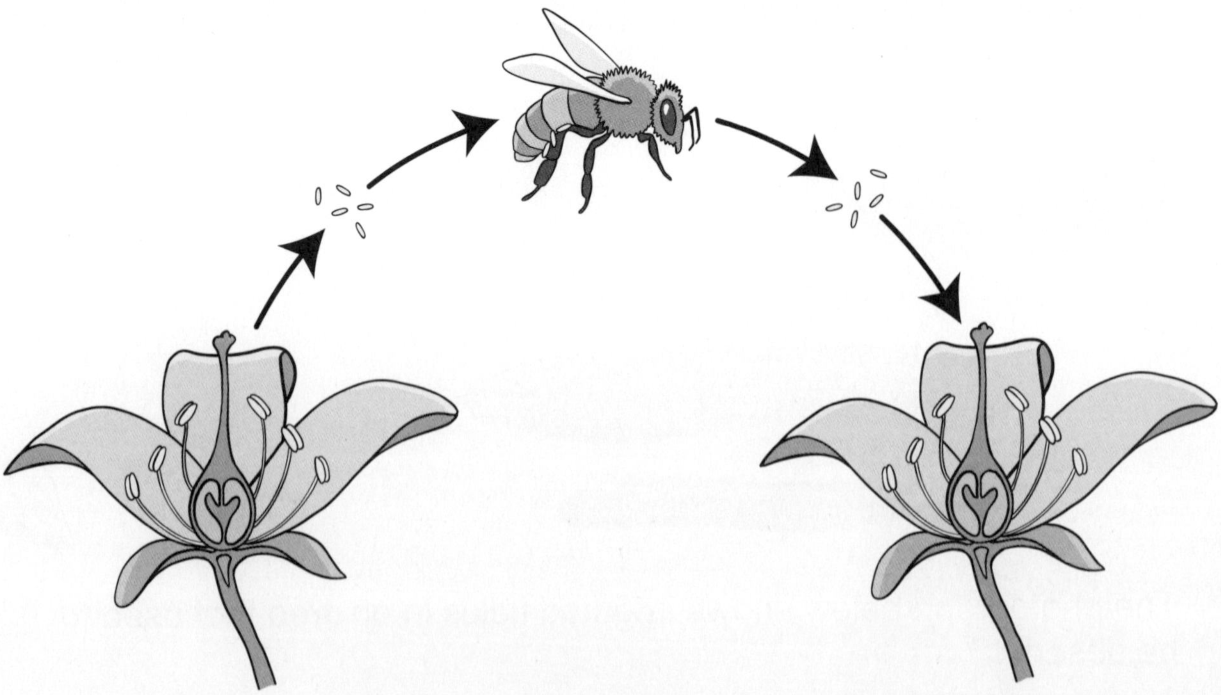

a What is the bee doing in the picture?

____

b Why is it important that the bee does this?

____

c What would happen if there were no more bees around to do this?

____

2 Circle the correct word to complete the sentence.

Bees drink **pollen** / **nectar** from flowers.

3 What is killing bees? Tick (✓) all the correct answers.

pesticides ☐ honey ☐ habitat loss ☐

beehives ☐ climate change ☐ sunlight ☐

4 Colour in the pictures that show how you can help bees.

Date: ____________________

Now look at and think about each of the *I can* statements.

1 Look at the pictures. Then answer the questions.

a Why can we say that the goat is a living thing?

____

b Why is the cardboard box non-living?

____

c Was the cardboard box once living? Explain your answer.

____

2 Look at the pictures. Then answer the questions.

Fossil A

Fossil B

a What types of fossils are shown above?

____

b How were these fossils formed?

____

____

____

Summative assessment 1

3 Look at the picture of a food chain. Then answer the questions.

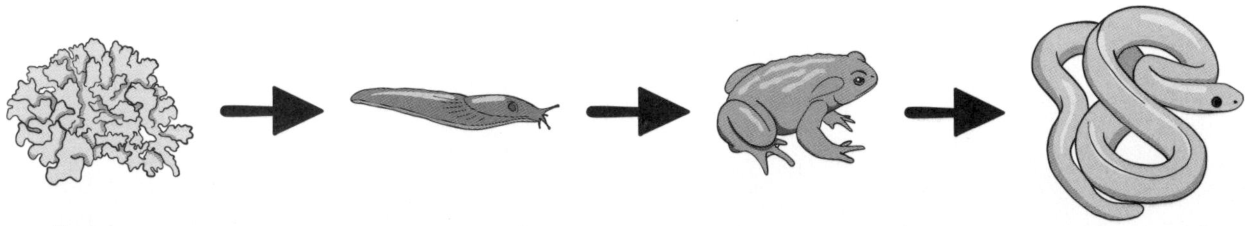

lichen great grey slug common toad grass snake

a What is the producer in the food chain?

____

b Name all the consumers in the food chain.

____

4 Look at these pictures from a food chain. Put the food chain in the correct order. Write the names and add the arrows.

stoat

blackberries

bank vole

____

5 Add labels to the picture of the plant. Use the words from the box.

flower	roots
leaf	stem

Summative assessment 1

6 Look at the pictures, then answer the questions.

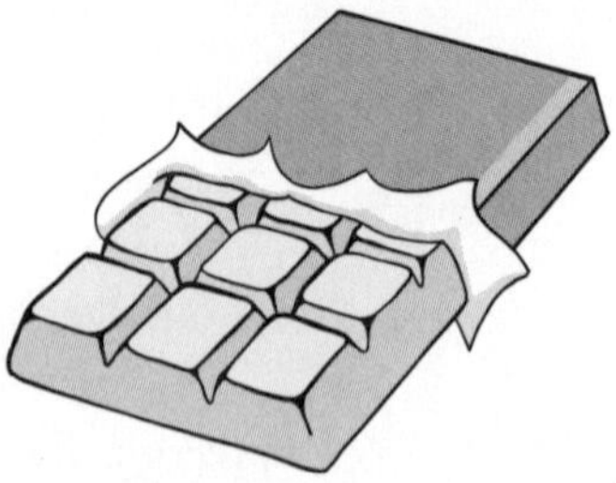

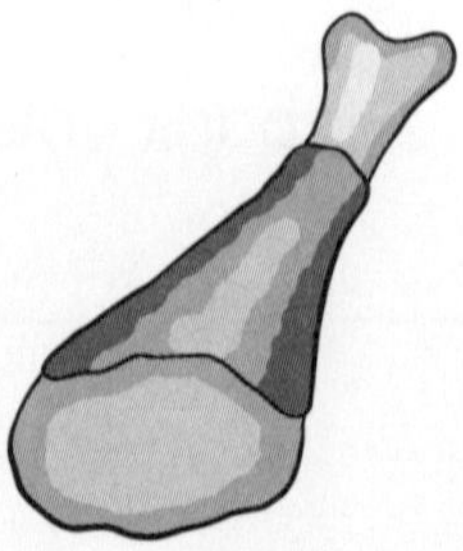

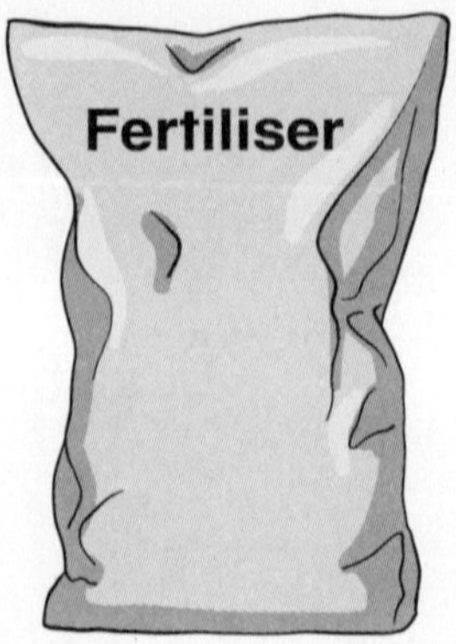

a Circle the things that plants need to grow and be healthy.

b Explain why plants need these things to grow.

____

____

____

____

c Which item above is used by farmers to help their crops grow well?

____

7 Class 3 conducted an investigation on the same type of plant.

water and sun | water but NO sun | sun but NO water

a What is being tested?

___

b What can you observe from the investigation?

___

___

8 Explain why bees are important.

___

___

9 What do bees drink from plants?

___

Date: ___

1 Look at the picture, then answer the questions.

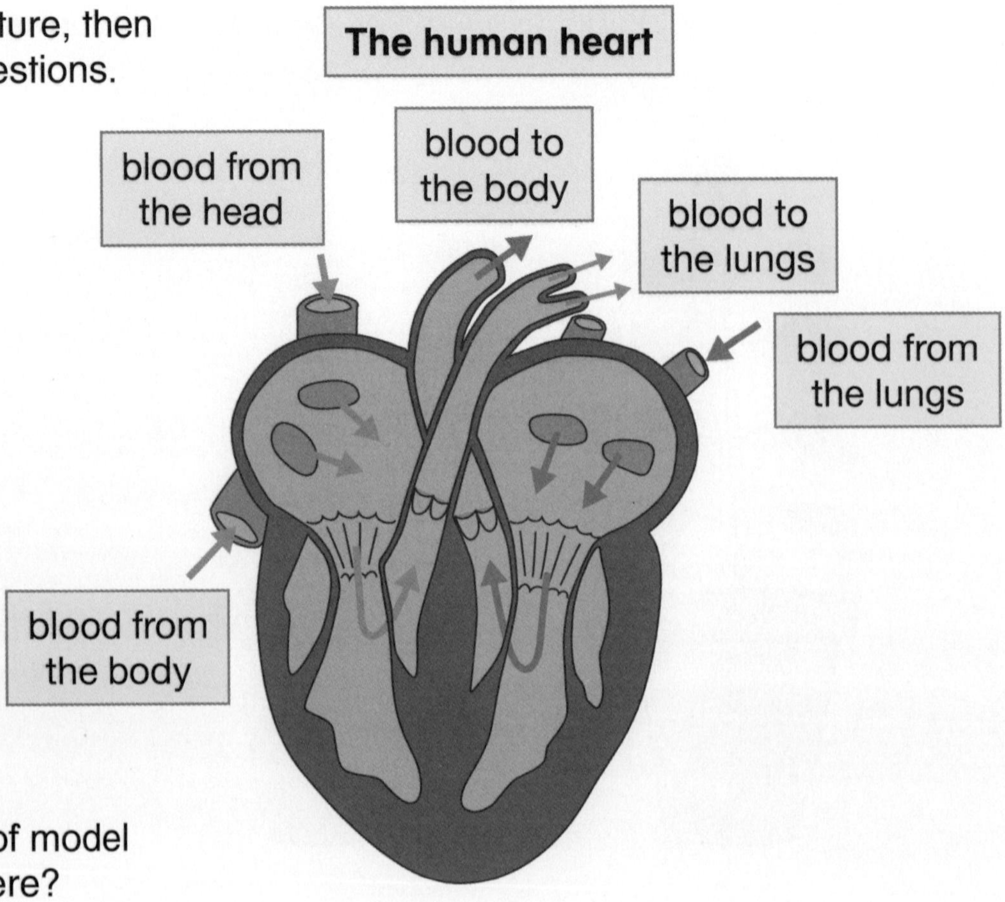

a What type of model is shown here?

____

b What is the model representing?

____

2 Circle the correct word to complete the sentences about the model in question **1**.

a The model is a **diagram / drawing** of a human heart.

b A human heart is about 12 cm long. This model is **smaller / bigger** than a real heart.

3 Look at this model of a human heart. Then answer the questions.

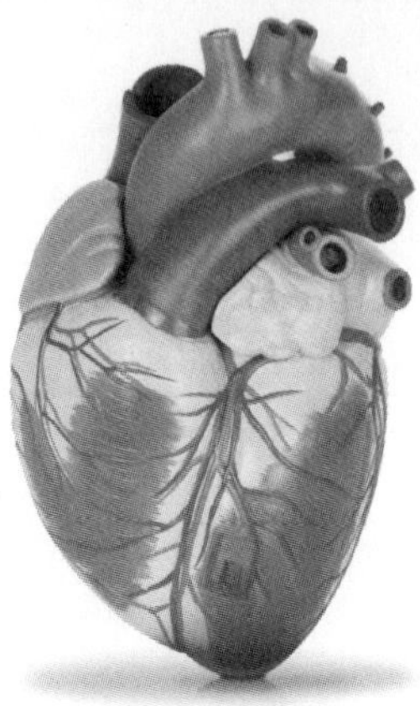

a What type of model is shown in the picture?

____

b How is this model of the heart better than the model in question **1**?

____

4 Decide whether the information about models below is suited for two-dimensional models, three-dimensional models or both. Tick (✓) the correct column.

Information	Two-dimensional model	Three-dimensional model
It helps us to understand science better.		
It is a diagram, picture or drawing.		
It can show something that is small a lot bigger.		
It is a representation of a real thing.		
It shows the actual shape of a real thing.		
It can show something that is big a lot smaller.		
It helps us to get a better understanding of what something looks like on the outside, inside and from all angles.		

Date: ____________________

Now look at and think about each of the *I can* statements.

1 Look at the pictures. Classify the animals by writing a word from the box under each animal.

insect	mammal	reptile	fish	bird	amphibian

a

____

b

____

c

____

d

____

e

____

f

____

2 Draw lines to match each animal to the correct classification and description.

Classification	Animal	Description
mammal		They have scales instead of fur, are cold-blooded and lay eggs instead of having live young.
reptile		They have fins instead of legs, are covered in scales and live in water.
insect		They have fur or hair, warm blood and give birth to live young, which they feed with milk.
fish		They have six legs, antennae on their heads and a hard outer shell covering their bodies.
bird		They have gills, but can also breathe air, often through their skin. They lay their eggs in water.
amphibian		They have two legs, feathers, a beak and wings. They are warm-blooded and lay eggs.

Now look at and think about each of the *I can* statements.

Date: ____________________

1 Look at the pictures.

a Work out the sequence of the life cycle. Number the boxes 1–4 to show the correct order.

b Use the words in the box to label each picture.

pupa egg butterfly caterpillar (larva)

c What changes take place in the life cycle of a butterfly?

d Does the egg look like the butterfly? Explain your answer.

2 Look at the pictures.

a Decide on the sequence of the human timeline. Number the boxes 1–7 to show the correct order.

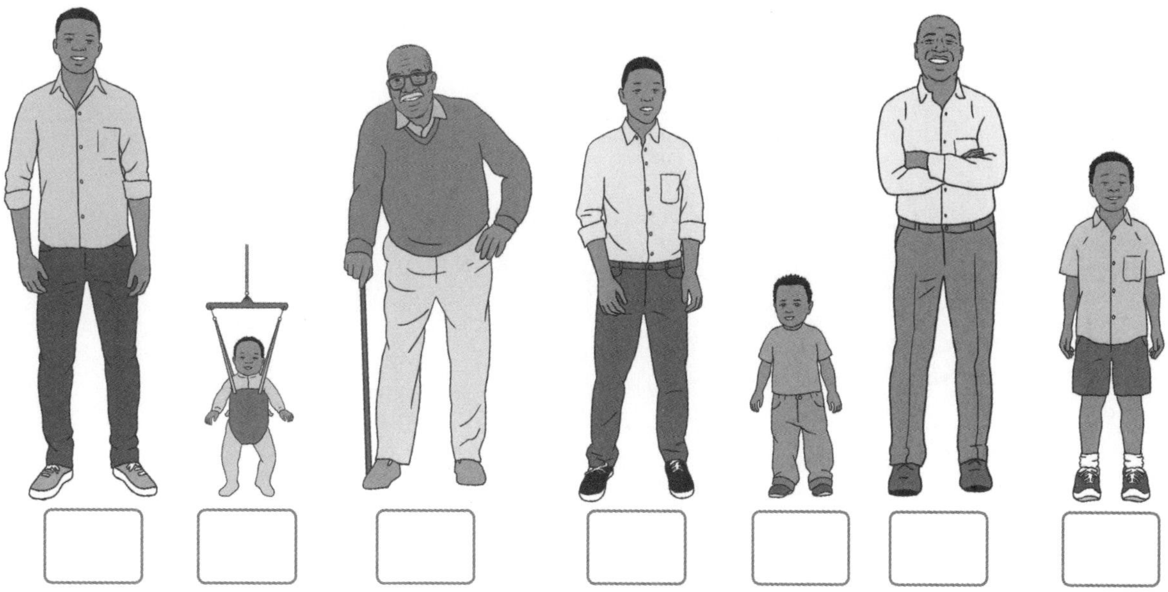

b Use the words in the box to identify the sequence of the human timeline.

teenager baby old man young adult toddler child adult

____

____

c What changes take place in the timeline of a human?

____

____

d Does the baby look like the old man? Explain your answer.

____

____

Now look at and think about each of the *I can* statements.

Date: ______________________

1 Use the words in the box to label the diagram of the brain below.

cerebrum cerebellum brainstem

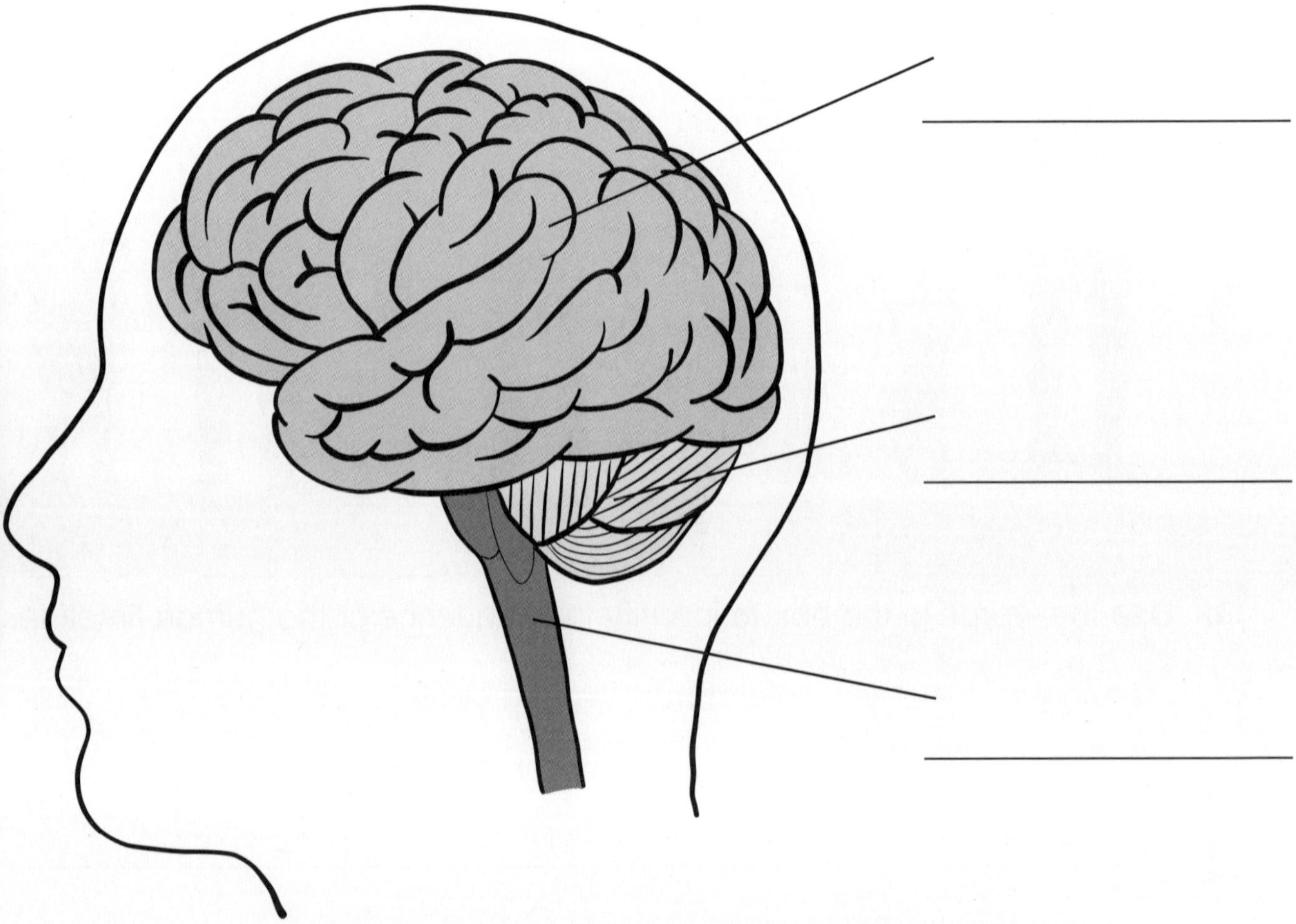

2 Draw lines to match each part of the brain with its function.

Part of the brain	Function
brainstem	The largest part of the human brain. It is folded to give it a bigger area. This helps it to store lots of information. It is where our memory is and where we work things out.
cerebrum	It helps us with movement and balance.
cerebellum	It controls heart rate, breathing, sleeping and eating.

3 Circle the correct words to complete the sentence.

The **heart** / **brain** controls our body systems, responds to our senses and allows us to think, **forget** / **remember** and feel emotions.

4 Look at the pictures. Then answer the questions.

a Why are the children not allowed to participate in these two activities without head protection?

____

____

b How is your brain protected from everyday falls or bumps?

____

____

5 Draw a picture of another activity you could do where head protection is needed.

Now look at and think about each of the *I can* statements.

Date: ____________________

1 Use the words in the box to complete the sentences.

pump	body	blood	minute	oxygen	muscle

The heart is a ____________________ that is about the size of our fist. It is made of ____________________, which helps it to do the job it needs to do. The heart never gets tired. It beats about 70 times each ____________________ for the whole of a person's life. The heart pumps ____________________ first to the lungs, then back to the heart and then around the rest of the ____________________. Our blood contains ____________________, which the rest of our body needs to function properly. It is the heart's job to pump the blood and oxygen to every part of our body.

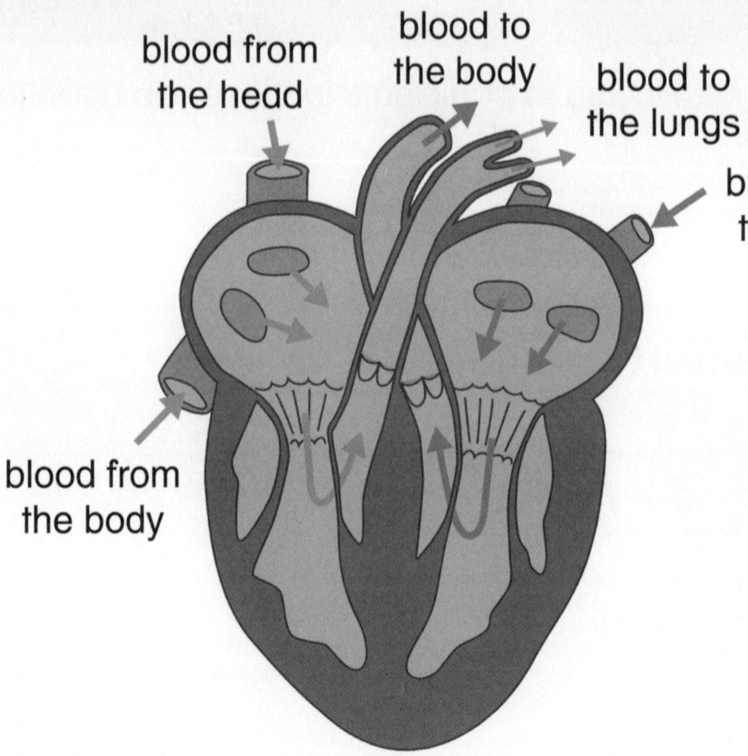

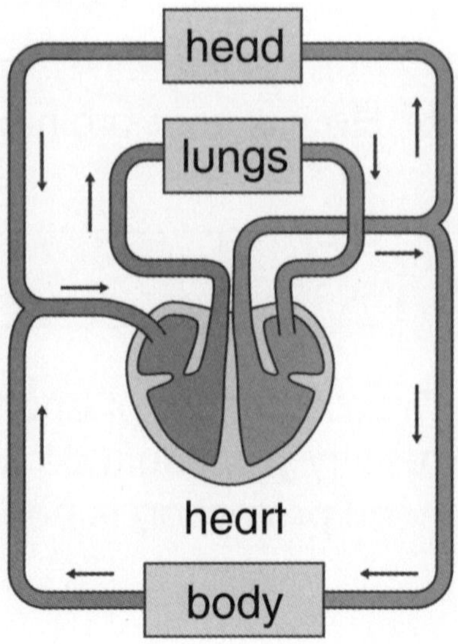

2 What do the arrows in the diagram in question **1** show?

____

3 Look at the information in the bar chart. Then answer the questions.

a Use the results from the bar chart to record the heart rate of each child.

Child	Heart rate
Nigel	
Anwar	
Sarah	
Aneena	
Moosa	

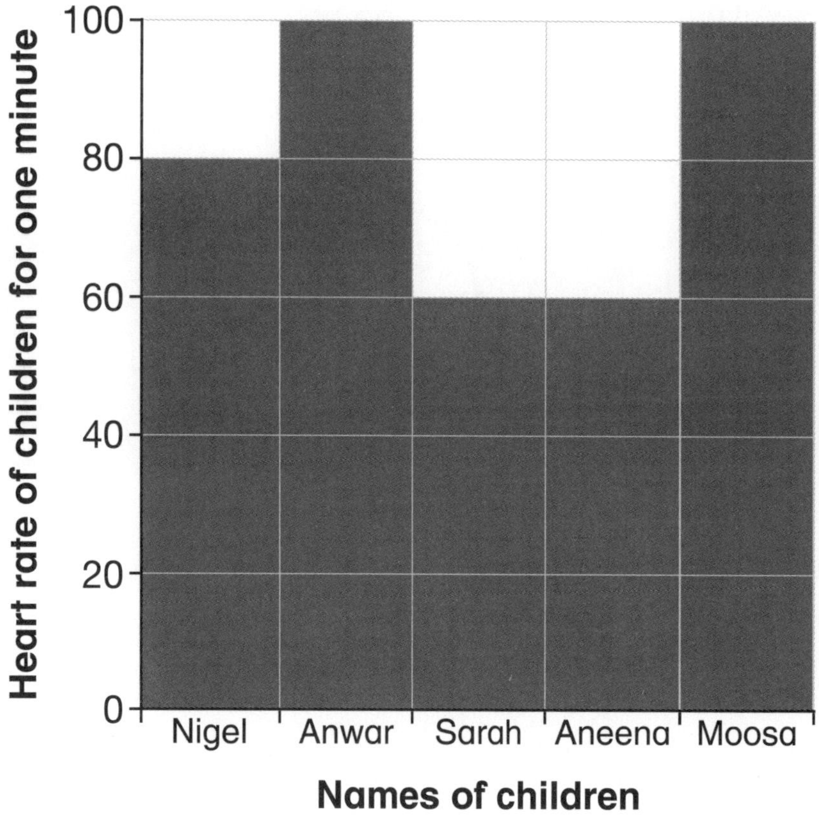

b Which children have the fastest heart rate?

____

4 The heart rates in question **3** were recorded when the children came back to class after play time.

a What do you think Anwar and Moosa were doing during play time?

____

b What do you think Sarah and Aneena were doing during play time?

____

c What helped you make your decision on the activities the children were doing during play time?

____

Date: ______________________

Now look at and think about each of the *I can* statements.

1 Use arrows to show the path that food takes through the human digestive system. Start at the mouth and end at the large intestine.

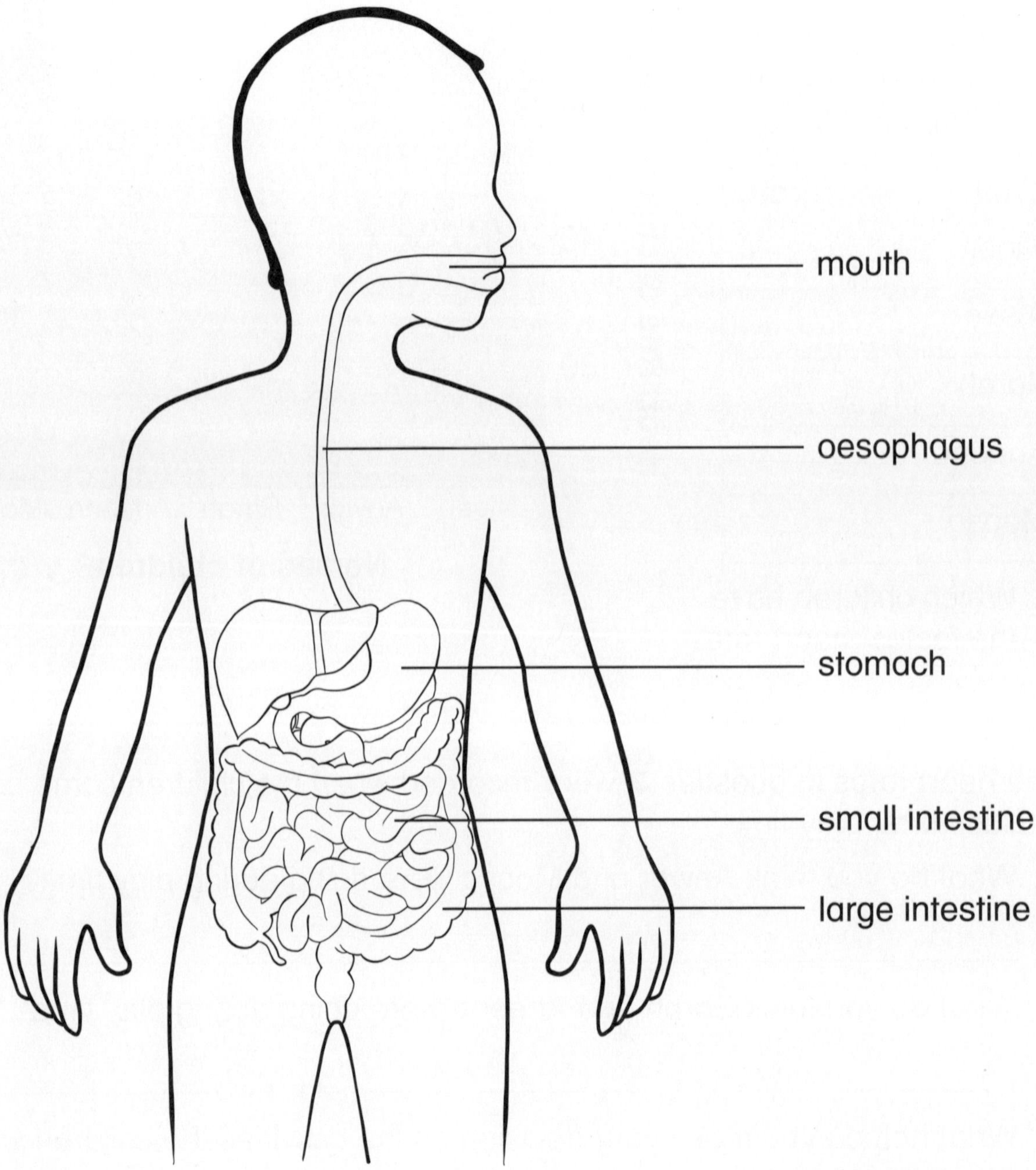

2 What is the name of the process where food gets broken down into smaller and smaller bits?

____

3 What happens to the food at every stage in the process you identified in question **2**?

Part of the body	What happens there?
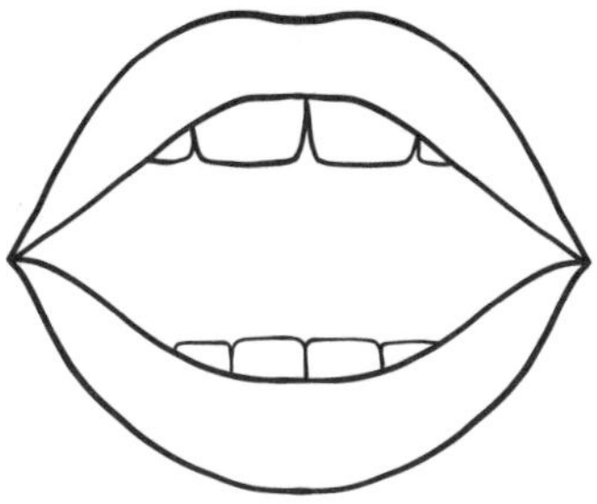 mouth	
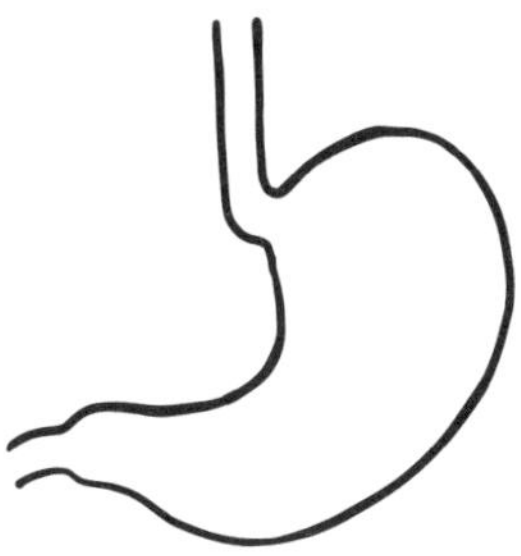 stomach	
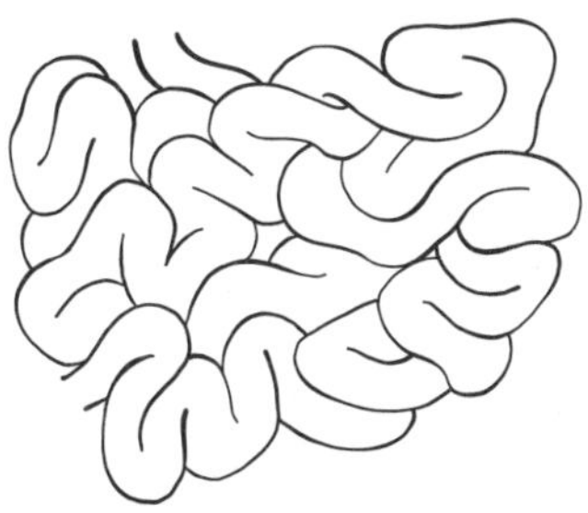 small intestine	
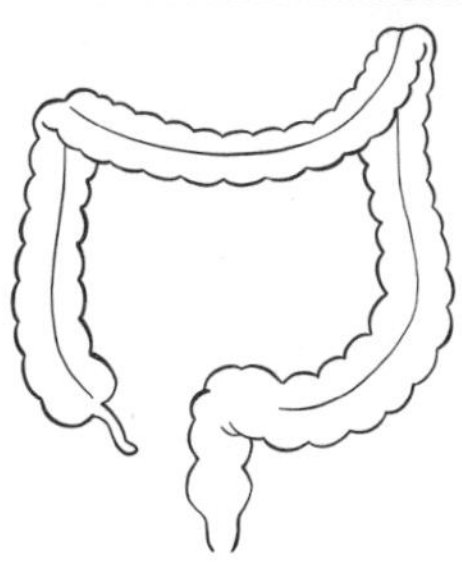 large intestine	

Date: ____________________

Now look at and think about each of the *I can* statements.

1 Look at the diagram.

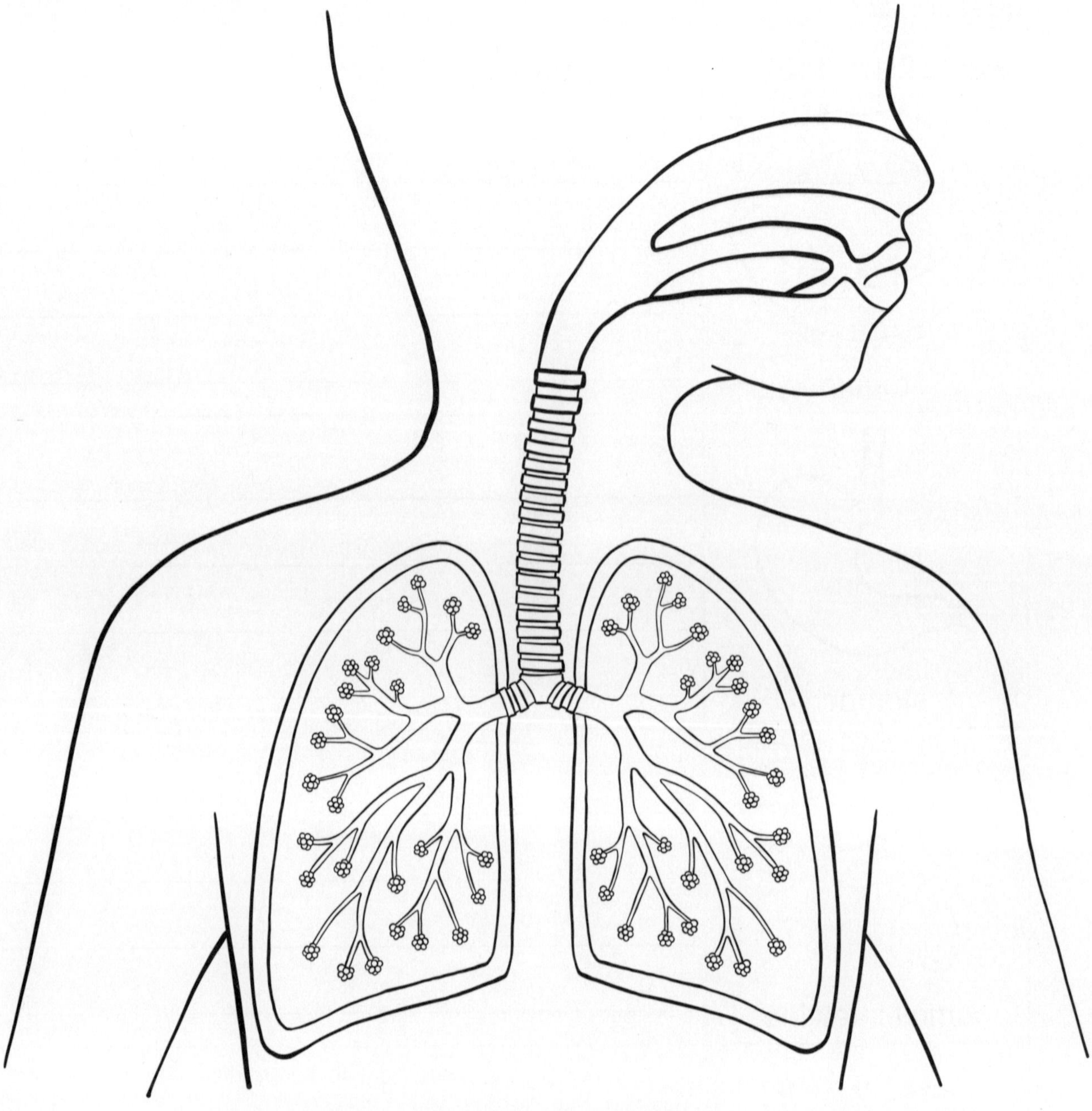

Follow these instructions to complete the diagram.

a Draw arrows in red to show how oxygen enters the body.

b Draw arrows in blue to show how carbon dioxide leaves the body.

c Label the diagram using the words in the box.

lungs	mouth	nose

2 Look at the diagrams, then answer the questions.

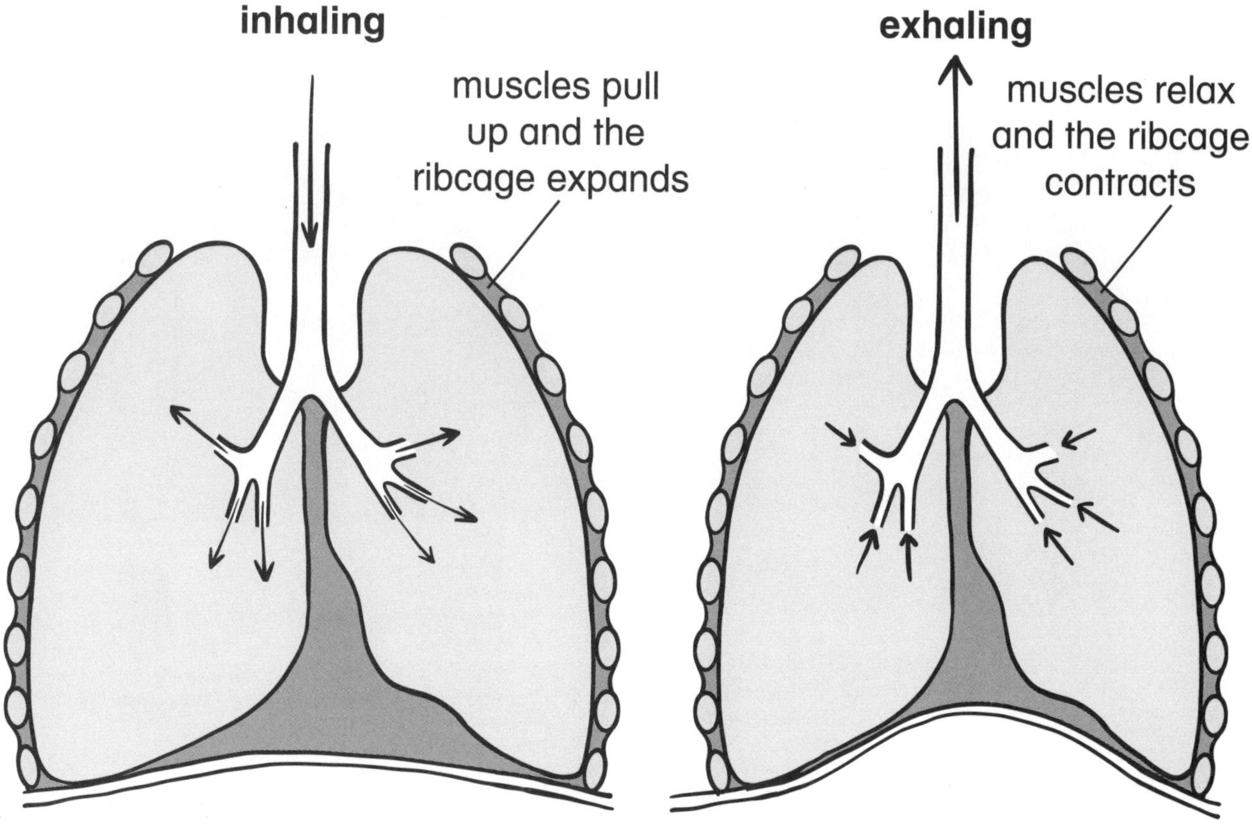

a What happens when you inhale?

____

b What happens when you exhale?

____

c What protects the lungs?

____

d What happens to your ribcage when you inhale?

____

e What happens to your ribcage when you exhale?

____

Date: __________

Now look at and think about each of the *I can* statements.

1 Look at the pictures. Which picture shows keyhole surgery?

____

Picture 1

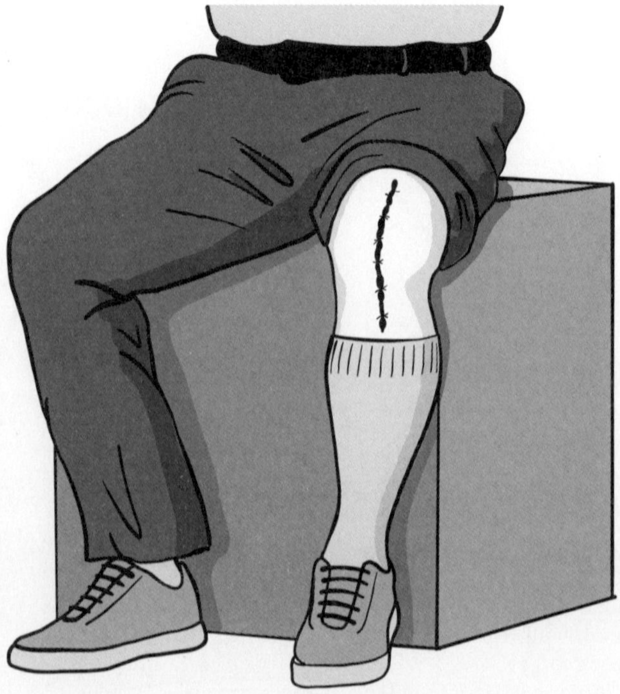

Picture 2

2 Explain how keyhole surgery is different from other types of surgery on a knee.

____

____

3 Give four reasons why keyhole surgery is an important development for cardiac surgery.

____

____

____

____

4 Name three different types of medical imaging that have helped surgeons.

- ______________________________
- ______________________________
- ______________________________

5 Draw lines to match the different types of scans to their uses for doctors.

Type of scan	Uses
MRI	This scan produces clear images of the inside of the body, including organs, such as the heart, so doctors and surgeons can see where they need to operate or if there is any disease inside a person's body.
CT	This scan creates an image of the inside of the body. Surgeons use this type of scan to help guide them during operations, such as heart surgery. They can see live images of the inside of the heart on a screen while they are operating on a patient.
ultrasound	

6 Circle the correct word to complete the sentence.

X-rays / **CT scans** / **MRI scans** have been used for many years to determine whether a bone is broken.

Now look at and think about each of the *I can* statements.

Date: ______________________

1 Is it a solid, a liquid or gas? Circle the correct word.

a It keeps its shape and takes up a definite space. **solid / liquid / gas**

b It can flow and has no fixed shape, so takes the shape of the container it is in. **solid / liquid / gas**

c It has no definite shape. It takes up all the space available and it can flow. **solid / liquid / gas**

2 Look at the pictures. Identify the solids, liquids and gases. Write **solid**, **liquid** or **gas** under each picture.

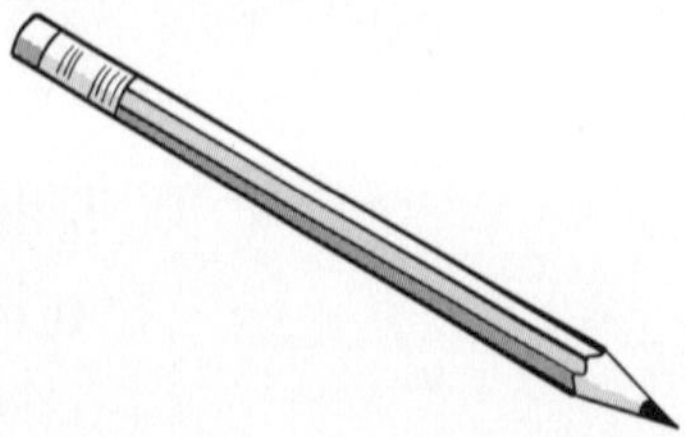

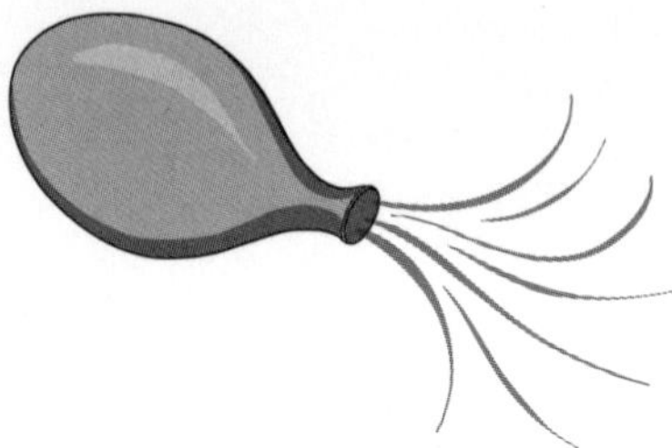

3 All the materials we use come from the Earth. Draw lines to match each material to an example and how it was formed.

Material from Earth	Example	How it was formed
natural gas	silver, aluminium, copper	Most come from the Earth's crust. They are found in rock, from which they can be extracted.
metals	crude oil, petrol	It is formed when layers of decomposing plants and animals are put under pressure and heat over millions of years.
oil	gas used for cooking	It comes from organic material which was quickly buried in mud at the bottom of ancient oceans and lakes. Over time, with pressure and heat, it changed into oil.

4 Which natural material is each arrow pointing to in the pictures? Write **metal**, **natural gas** or **oil** under each picture.

______________	______________	______________

Date: ______________

Now look at and think about each of the *I can* statements.

1 Look at the picture. It shows an investigation to find out which solids float in water.

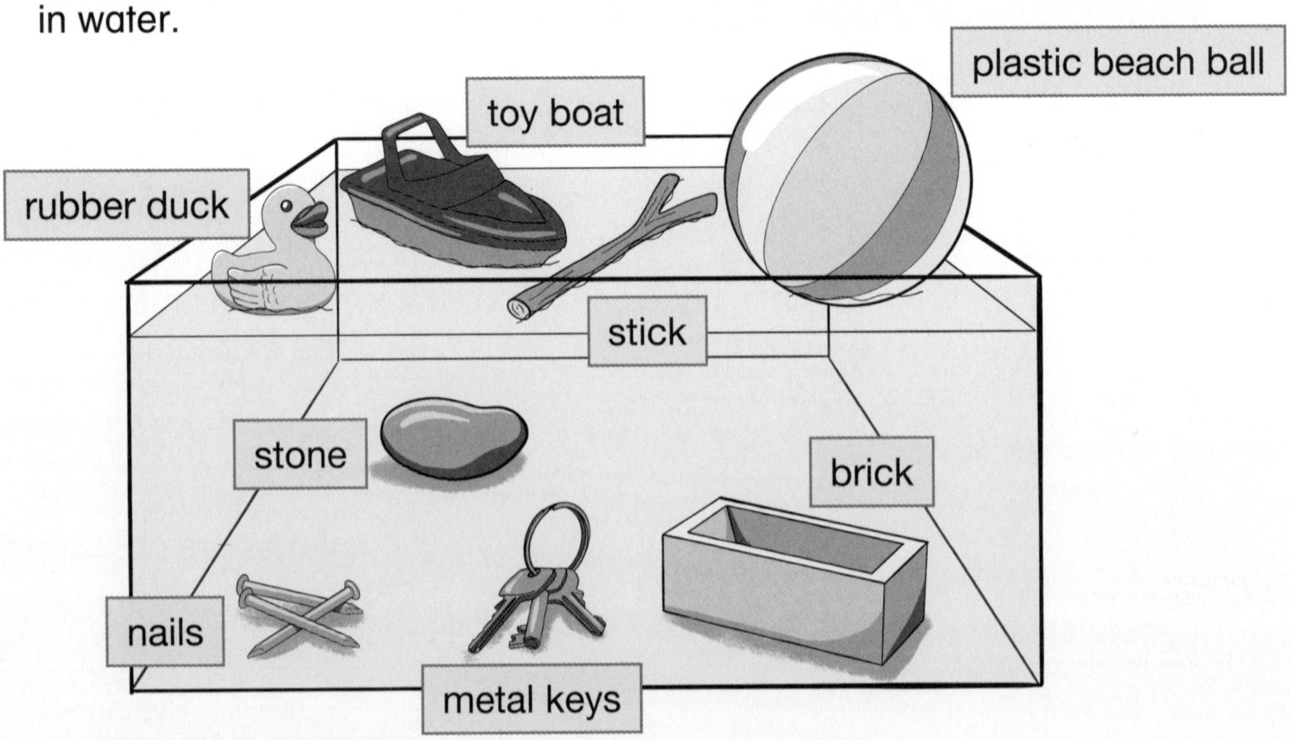

a What can you observe from the investigation?

____

b List the objects that floated on the water.

____

c List the objects that sank to the bottom of the container of water.

____

d What do you notice about the properties of the solids that floated and the ones that sank? How are they different?

____

____

____

e What safety precautions would you need to follow when conducting this investigation?

____

____

2 Is the solid malleable or brittle? Tick (✓) the correct column.

Solid	malleable	brittle
ceramic dinner plate		
rope		
metal paper clip		
glass vase		
copper wire		

3 Does the liquid pour quickly or slowly? Tick (✓) the correct column.

Liquid	Pours quickly	Pours slowly
milk		
potato soup		
fruit juice		
liquid paint		
olive oil		

4 Circle the correct word to complete each sentence.

a Liquids flow and cannot be easily **compressed** / **dispersed**.

b Solids have a definite shape that is **not easy** / **easy** to change.

c Some solids are brittle and **break** / **squash** easily.

d Liquids **always** / **sometimes** take the shape of the container they are in.

Date: ______________________

Now look at and think about each of the *I can* statements.

1 Look at the pictures. Is Picture 1 or Picture 2 a mixture? Explain your answer.

Picture 1: bowl of nuts

Picture 2: pancake batter

2 Look at the pictures below. How would you separate each mixture? Write your answer in the second column.

Mixture	How would you separate it?
a 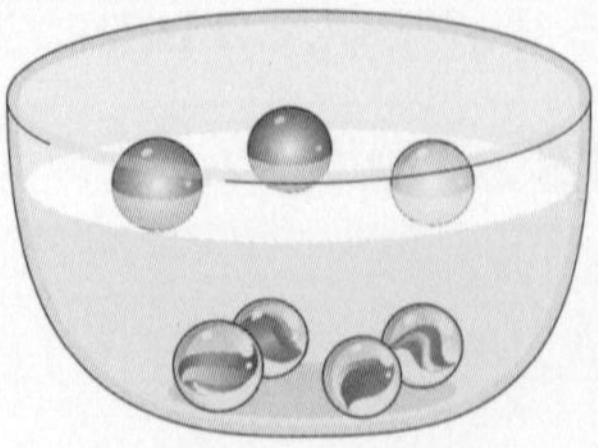	
b 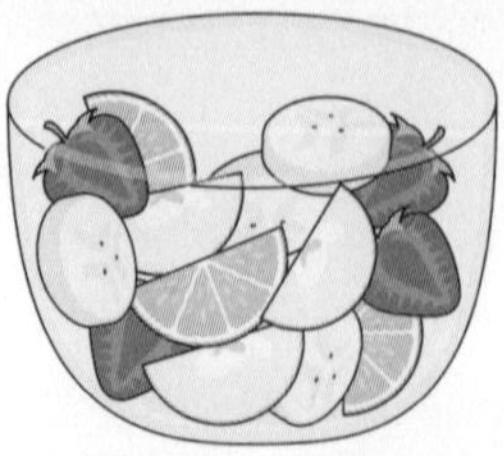	
c 	
d 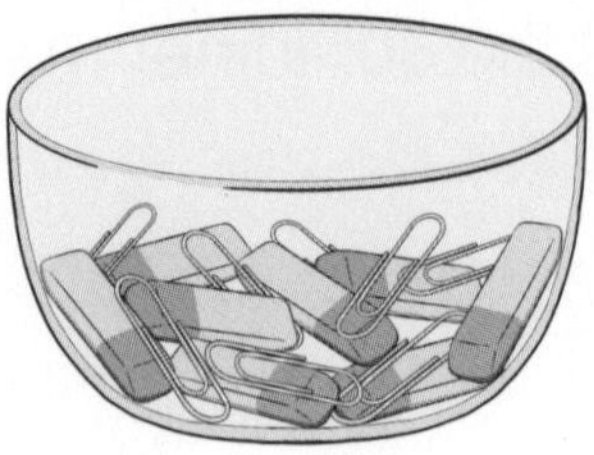	
e 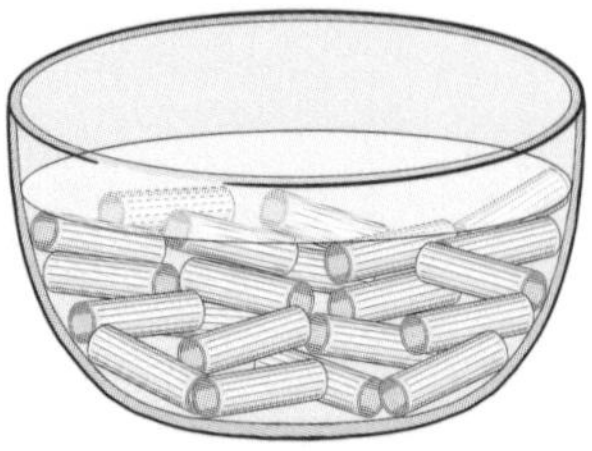	

3 Describe how you can separate a mixture of coffee grains and water.

____

____

____

4 Draw and label an example of a mixture you can separate using each of the methods below.

Method of separation	Example of mixture to separate
sieving	
filtering	
using a magnet	

Now look at and think about each of the *I can* statements.

Date: ______________________

1 Ground was sieved with a special mesh sieve. Which sieve will not allow the different solids to filter through it? Draw lines to match each sieve to the type of ground it will not let through.

Type of sieve	Solid mixture

a

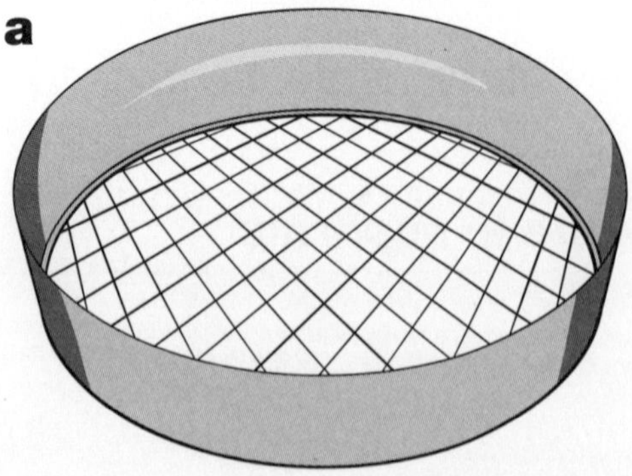

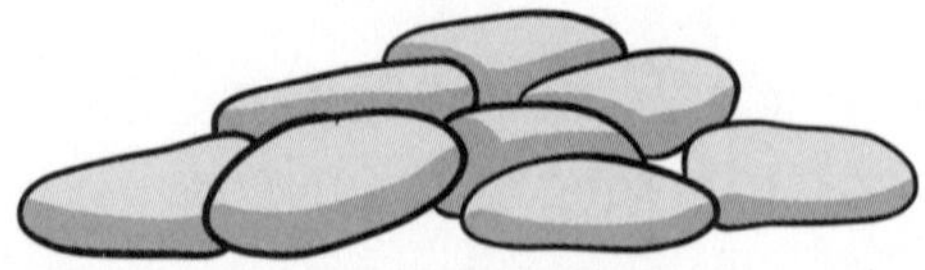

b

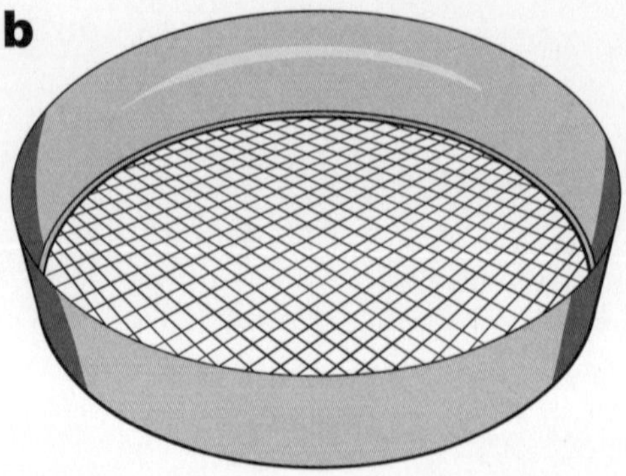

c

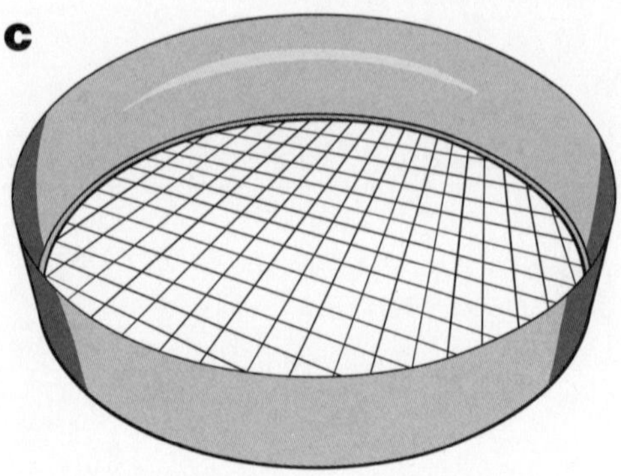

d

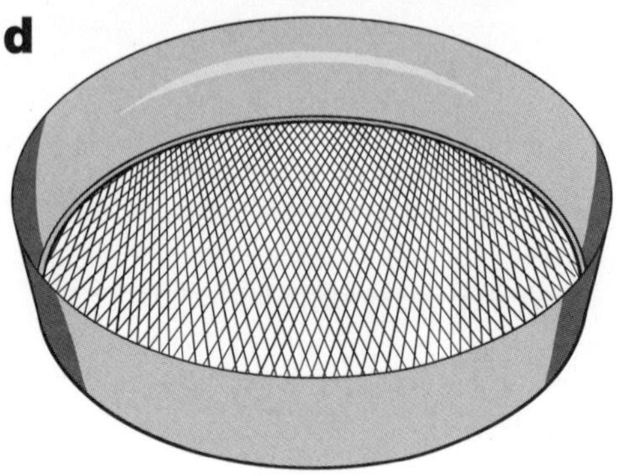

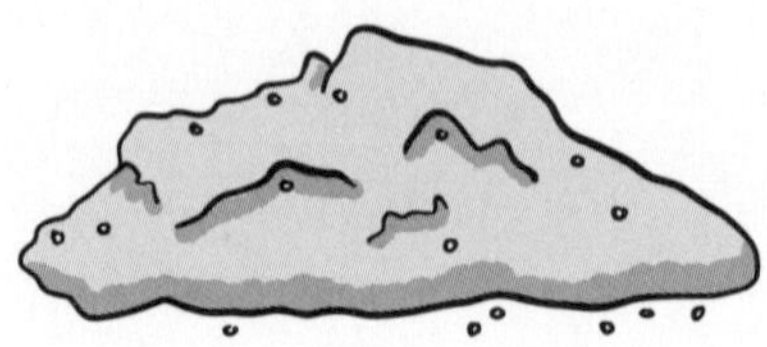

2 Look at the picture. Why would a sieve not work to separate this mixture of solids?

____

____

3 Look at the picture.

a What is being used to separate iron and steel from aluminium?

____

____

b Why is this method used to separate the metals and not a sieve?

____

____

4 A necklace was dropped into a sandpit while children were playing in it. Explain the sorting method you would use to find the necklace.

____

____

Now look at and think about each of the *I can* statements.

Date: ______________________

1 Look at the pictures. Choose which method you would use to separate the solid from the water.

Solid and water mixture	Method to separate them
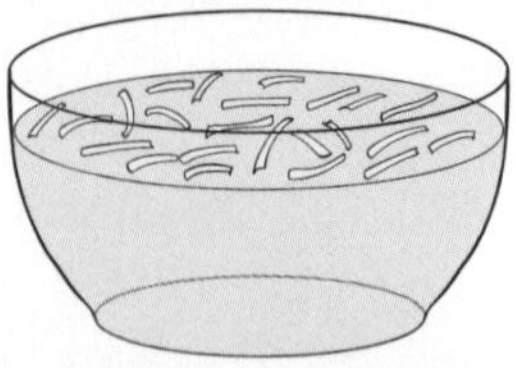 grated cheese and water	filter paper / colander
 pepper and water	filter paper / colander
 coffee grains and water	filter paper / colander
 rice and water	filter paper / colander
 sand and water	filter paper / colander

2 Think about your answers to question **1**. Why does a colander work for some mixtures but others need filter paper to separate them?

___

___

3 Look at the pictures. Circle the solids that will dissolve in water.

salt

jelly powder

sand

tea leaves

sawdust

bath salts

4 Which solids in question **3** change the colour of water?

____

5 When a solid has dissolved in water, will the water be clear or cloudy and difficult to see through?

____

6 Explain how to separate a dissolved solid from water.

____

____

Date: ____________________

Now look at and think about each of the *I can* statements.

1 Look at the pictures. Classify each animal into one of the six main groups.

______________________ ______________________

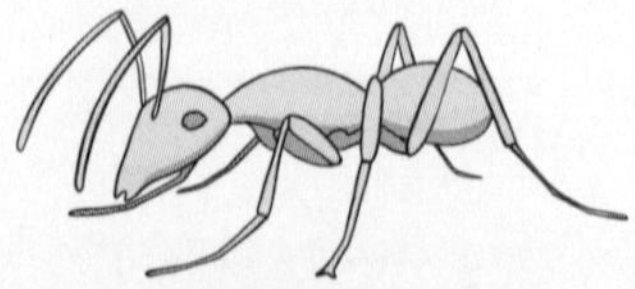

______________________ ______________________ ______________________

______________________ ______________________

2 Name one common feature of the animals in these groups.

a mammal: ______________________

b reptile: ______________________

c amphibian: ______________________

d insect: ______________________

e bird: ______________________

f fish: ______________________

3 Look at the diagram on the next page. How is the life cycle of a bee different from the timeline of a human?

____

____

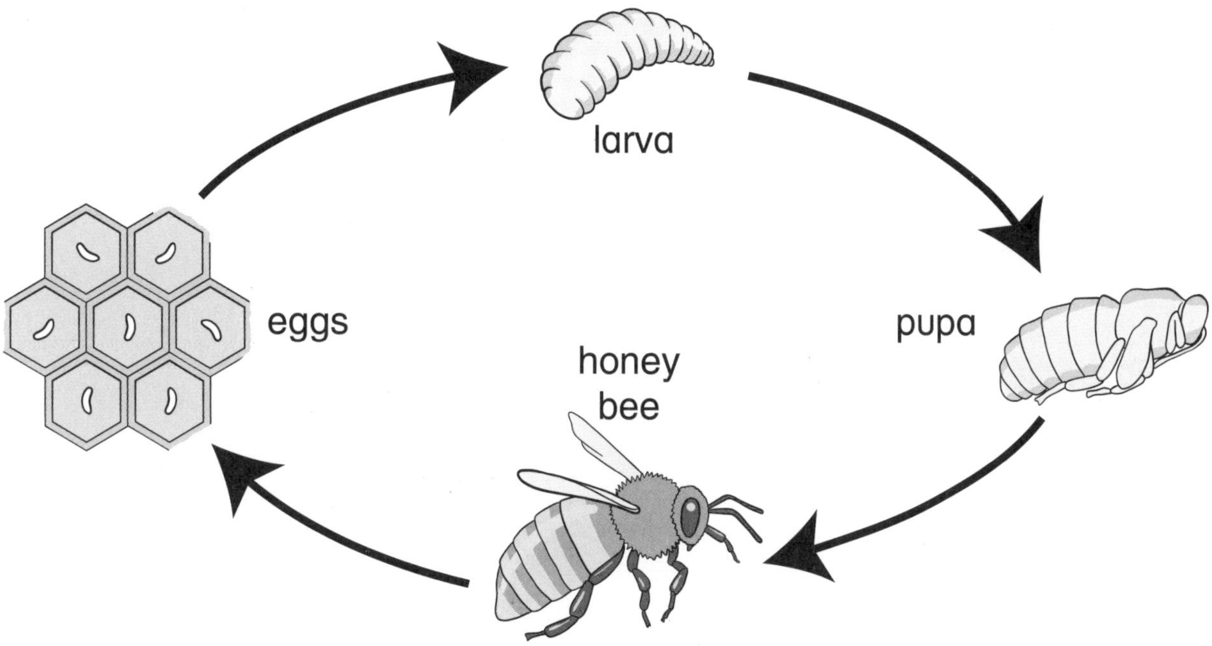

4 What is the importance of each part labelled on the diagram of the human body below?

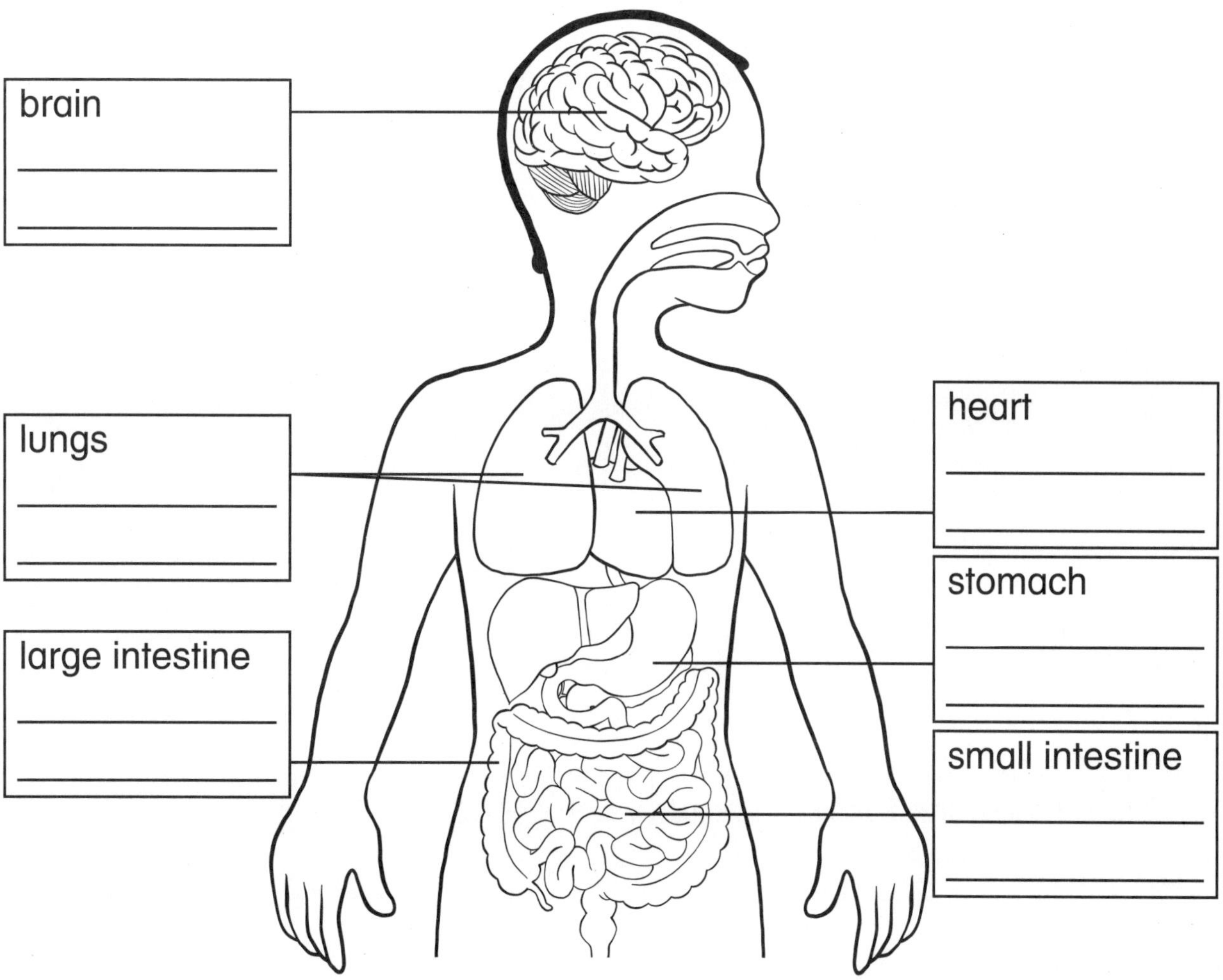

Summative assessment 2

5 Sort the objects below into solids, liquids or gases by writing them in the first row of the chart. Then describe their special properties in the second row.

	solids	**liquids**	**gases**
Examples			
Properties			

scissors

gas from cooker

rocks

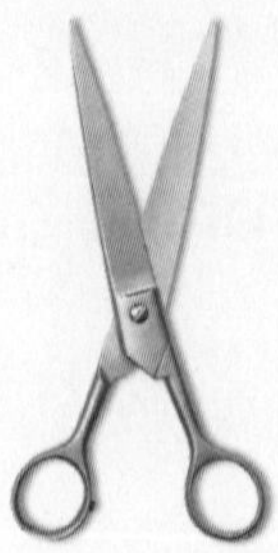

air in balloons

ice

juice

6 Where are oil, natural gas and metals found?

____

7 Explain how the following natural resources are useful.

a oil: ____________________________________

b natural gas: ______________________________

c metals: _________________________________

8 Look at the pictures of the different states of water. Then answer the questions.

__________ __________ __________

a Label each state of water as a **solid**, **liquid** or **gas**.

b What are the properties of solid water?

____

c What are the properties of liquid water?

____

9 Describe the sorting method that you would use to separate these solids from water.

a beans: __________

b washing powder: __________

c cooked pasta: __________

d sand: __________

10 a What happens when sugar is mixed with water? __________

b Explain how you can separate sugar that has been dissolved in water.

____

____

11 When would you use a magnet to separate a mixture?

____

Date: __________

1 Look at the picture. It shows an investigation to see whether different road surfaces change the speed of a car.

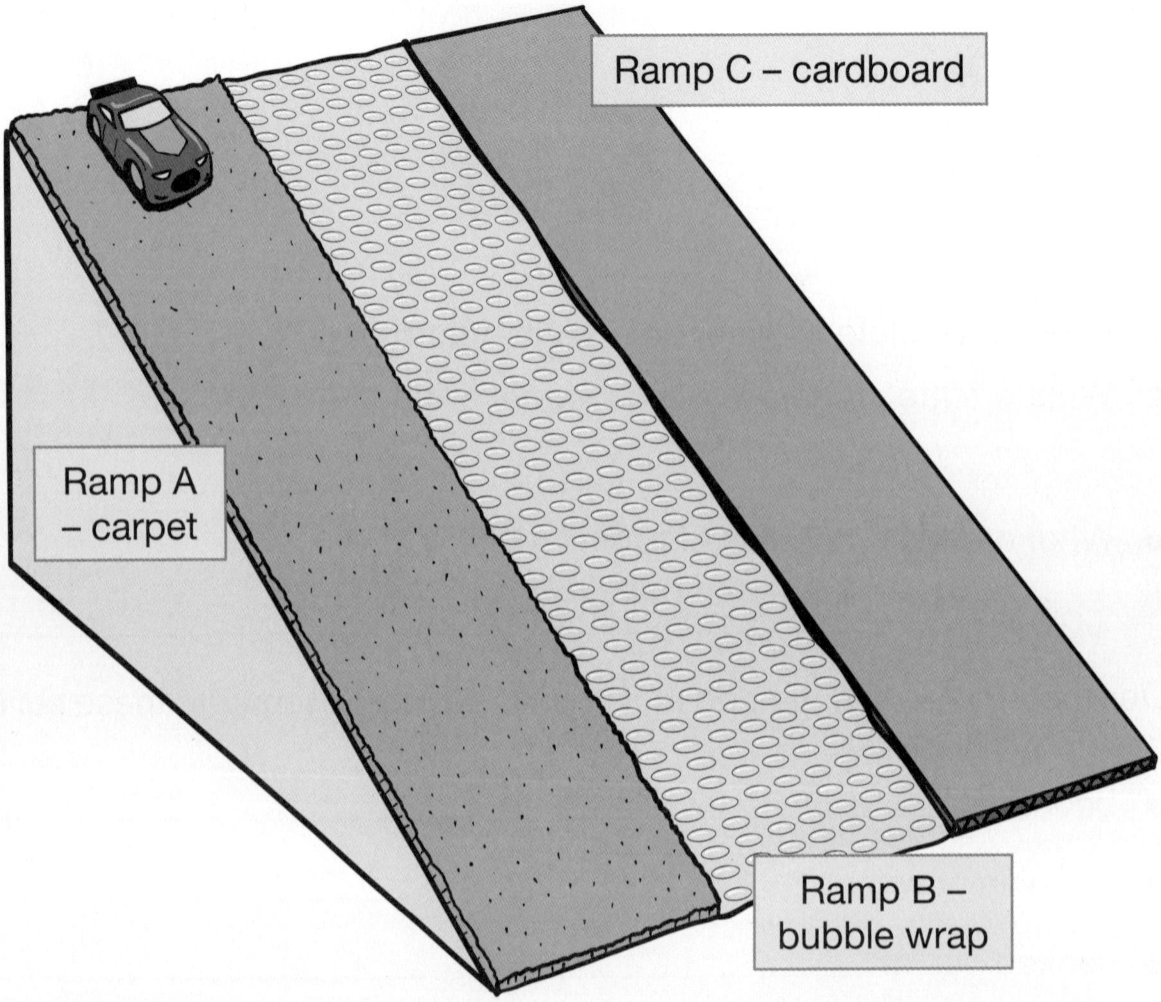

a What is kept the same in this investigation to make it a fair test?

____

b Why is it important to keep these things the same?

____

c What changes in each test?

____

2 The results of the car speed investigation in question **1** are recorded in the table.

Ramp	Speed in seconds
A	15
B	10
C	5

a Use the information from the table to complete the bar chart on the next page.

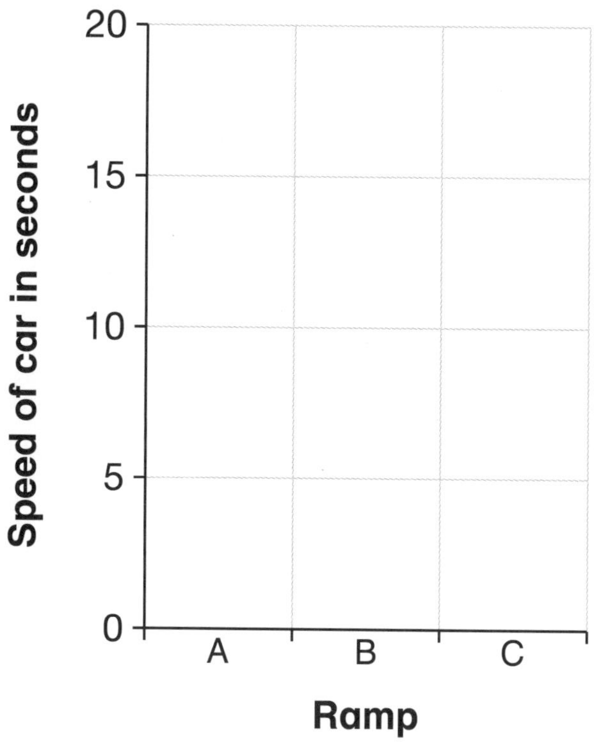

b Why do you think the bar chart looks like this?

___

c Which surface made the car move the slowest?

___

d Which surface made the car move the fastest?

___

e Can you identify a pattern in the results?

___

___

Date: ___

Now look at and think about each of the *I can* statements.

1 What is the name of the force that makes apples fall down towards the ground and not go up into the sky?

____

2 Circle the correct words to complete the sentence.

Gravity is a natural force which **pushes** / **pulls** everything towards the **centre** / **outside** of the Earth in a **downwards** / **upwards** direction.

3 Look at the picture. Then answer the questions.

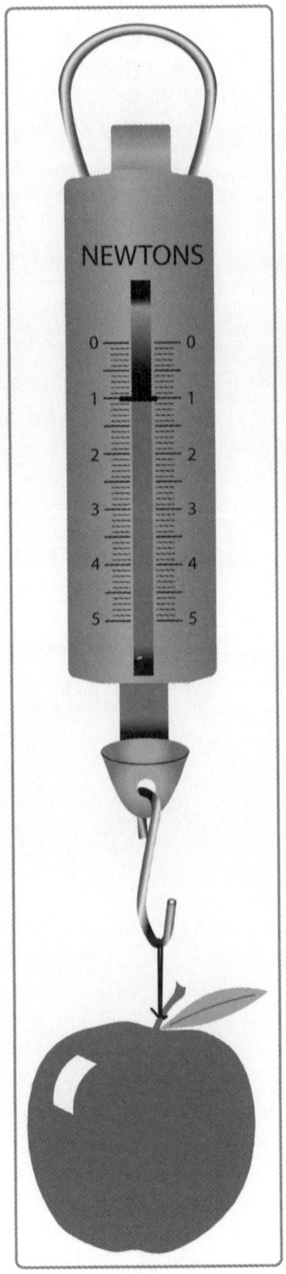

a What is the name of the equipment used to measure how strong a force is?

____

b How many newtons are shown in the picture?

____

c What is inside a force meter to measure the force?

____

4 Look at the picture of different objects being weighed on the next page.

a Record the measurements of the forces in the chart below.

Object	Measurement
pencil	
brick	
orange	
saucepan	

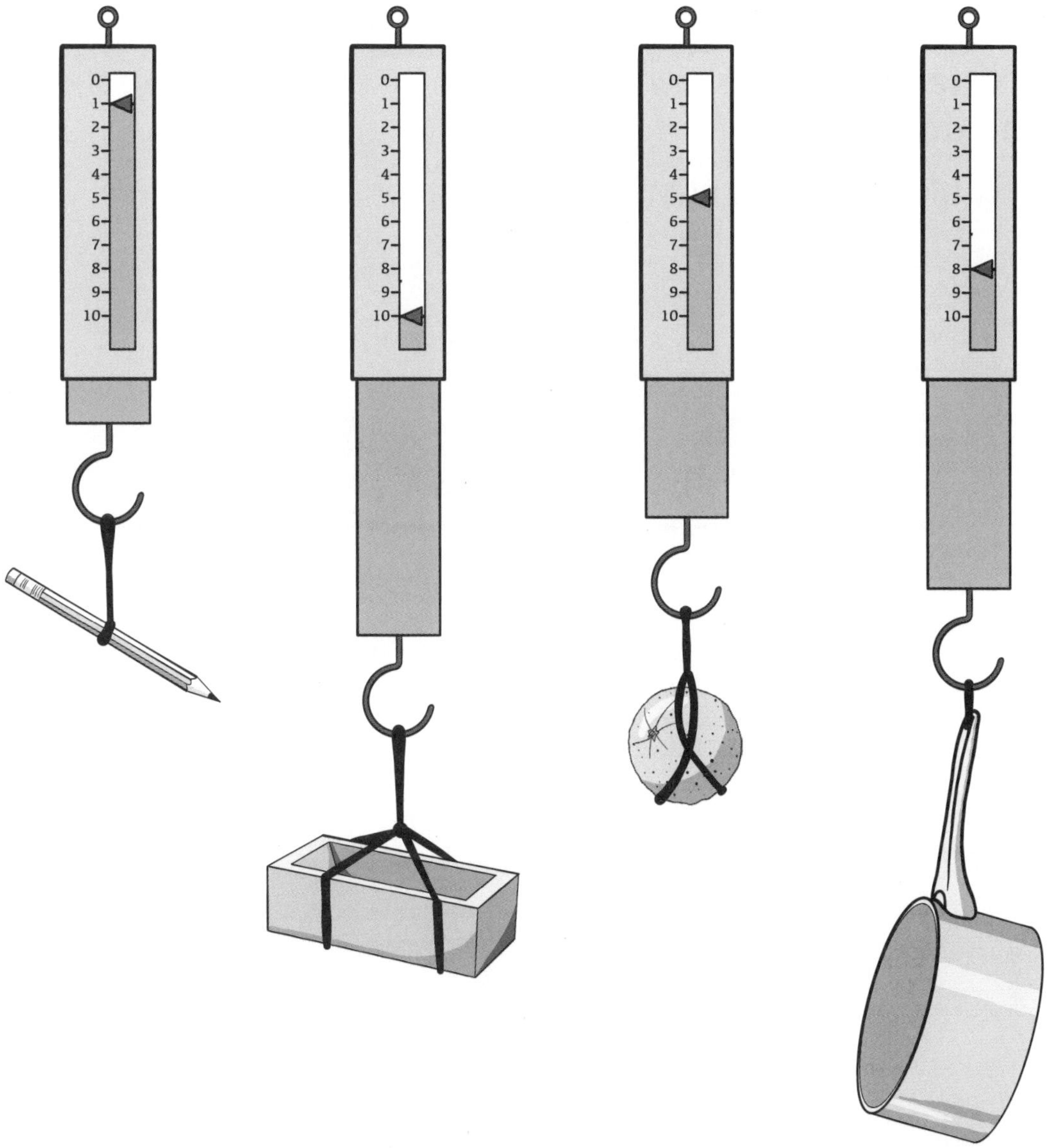

b Which weighs more, the pencil or the orange?

____

c Which weighs more, the brick or the saucepan?

____

d Which weighs more, the orange or the saucepan?

____

Now look at and think about each of the *I can* statements.

Date: ____________________

1 Look at the pictures. Then answer the questions.

a Draw red arrows to show the direction in which each object or person is moving.

b Draw blue arrows to show the direction of the friction force.

2 Circle the correct words or phrases to complete the sentences.

a Friction is a force that **speeds things up** / **slows things down**.

b Friction acts in the **opposite** / **same** direction to the movement.

3 Look at the pictures. Which footwear is better suited to playing a rugby match on grass? Explain your answer.

boots with studs

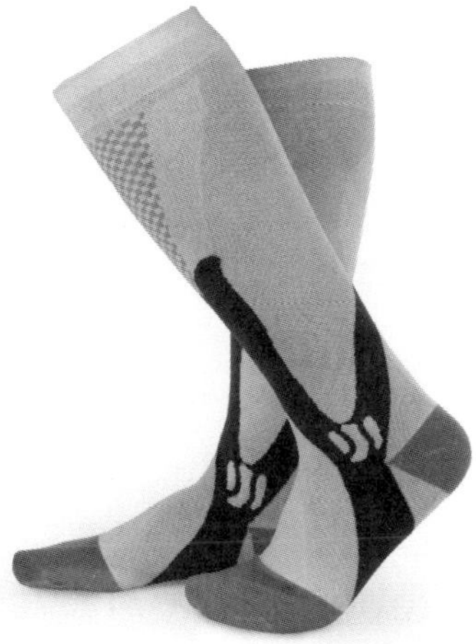

long sports socks

____

____

4 Look at the pictures. Which tyre is best suited for riding on a gravel road? Explain your answer.

mountain bike tyre

smooth, road tyre

____

____

Now look at and think about each of the *I can* statements.

Date: ____________________

1 Look at pictures **A** and **B**. Then answer the questions.

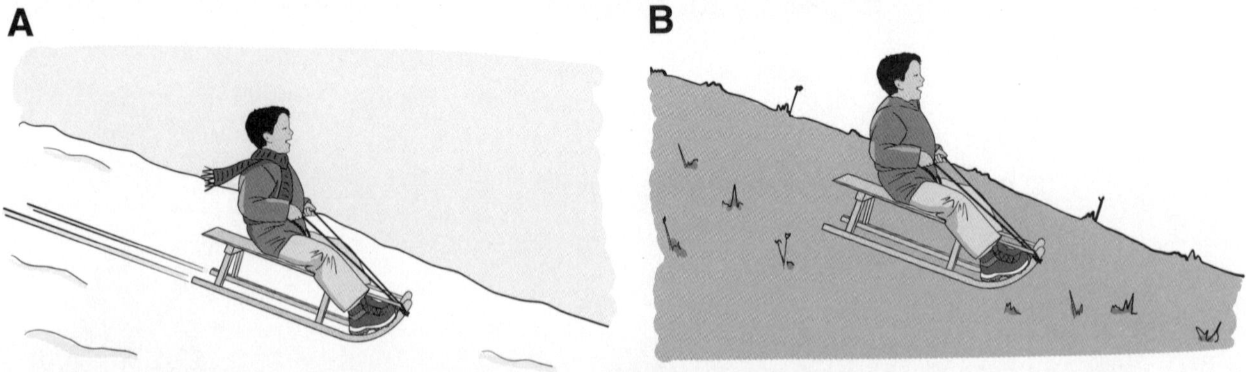

a Do the pictures show a fair test to see which surface allows the toboggan to move faster? Explain your answer.

____

b Which surface allowed the boy to move faster on the toboggan? Explain your answer.

____

____

c Which picture shows a resistance in movement?

____

2 Look at pictures **A** and **B**. Both boxes need to slide across the floor. Which box will slide more easily? Explain your answer.

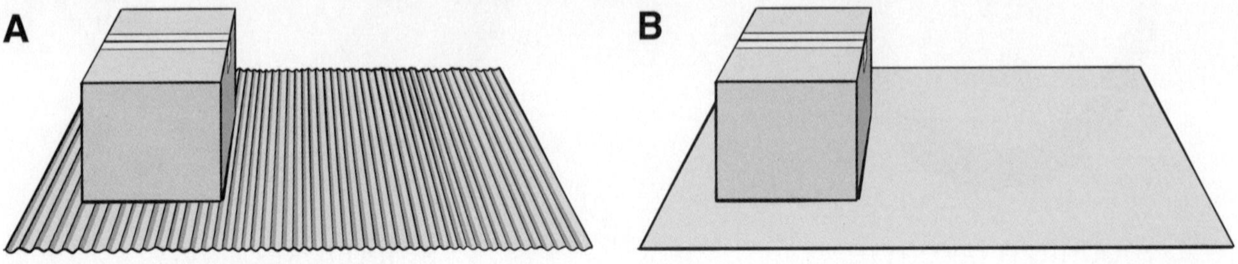

____

____

3 Look at the pictures.

a On which surfaces is it easy to slide? Write E next to the pictures.

b What makes the other surfaces difficult to slide on? Explain the reasons why.

____

____

____

____

Now look at and think about each of the *I can* statements.

Date: ______________

1 Look at the pictures. Is the object transparent, translucent or opaque? Tick (✓) the correct column.

	transparent	translucent	opaque

2 Complete the sentences by choosing the correct words from the box.

Transparent	Translucent	Opaque	light

________________ materials let ________________ through them, so objects can be seen on the other side. In some places we only want to let a little light in. ________________ materials are used for these objects. ________________ materials do not let any light through.

3 What is a shadow? Explain your answer.

____

4 What determines the size of a shadow?

____

5 Look at the pictures. Then answer the questions.

a What is the light source in the pictures?

____

b What causes the shadow to form in each picture?

____

c Write under each picture whether the shadow is **long** or **short**.

Now look at and think about each of the *I can* statements.

Date: ______________________

1 Will the magnets attract or repel each other? Write **attract** or **repel** under each pair of magnets.

a

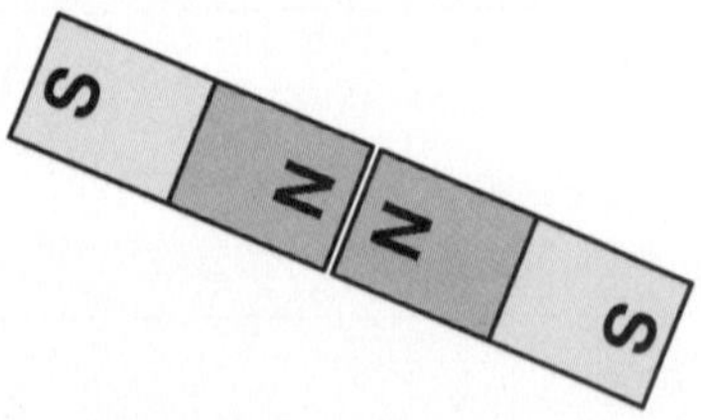

____

b

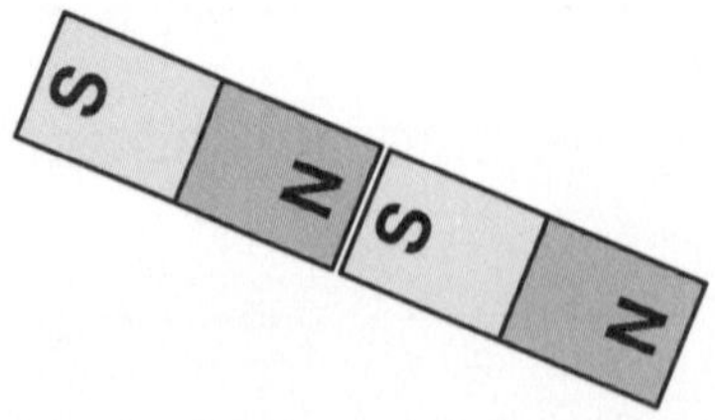

____

c

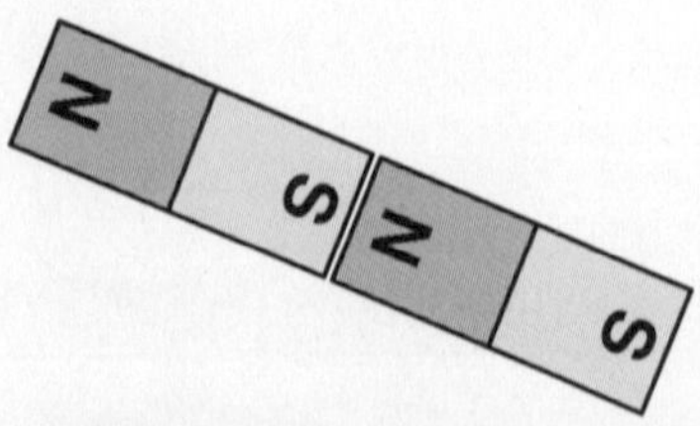

____

d

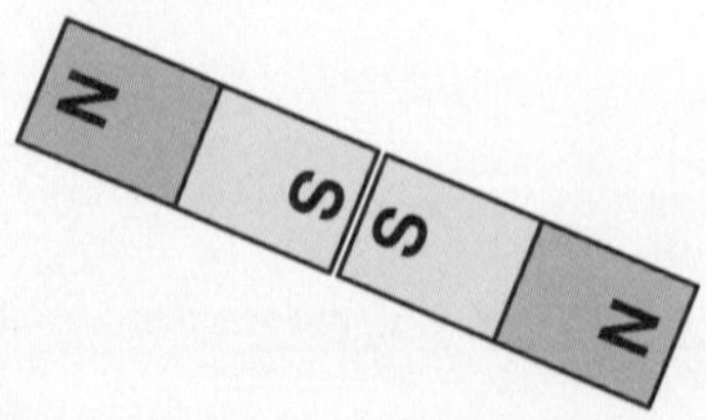

____

e

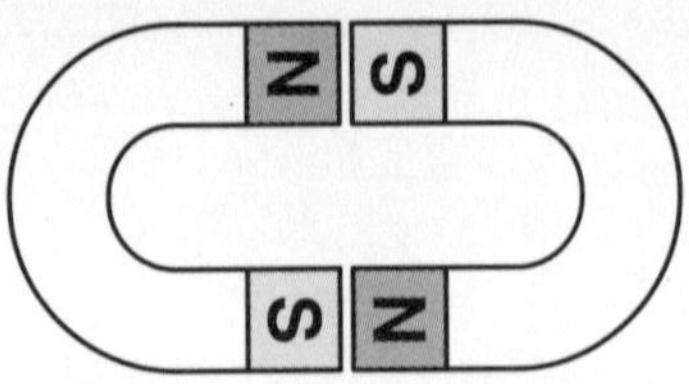

____

2 What do the letters 'S' and 'N' on each magnet stand for?

____

3 What do magnets attract?

____

4 Look at the pictures. Are the materials magnetic or non-magnetic? Write **magnetic** or **non-magnetic** under each picture.

____

____

____

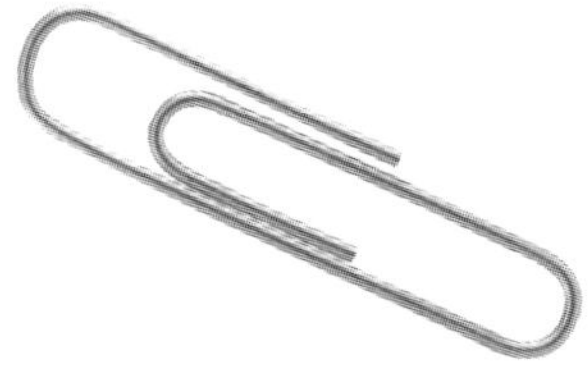

________________ ________________ ________________

________________ ________________ ________________

________________ ________________ ________________

________________ ________________

5 Look at the picture. What is the name of the object?

____

6 Circle the correct words to complete the sentences.

a The compass needle is a magnet that is **attracted** / **repelled** to **true** / **magnetic** north.

b The compass needle **always** / **sometimes** points towards magnetic north.

c True north is 500 kilometres **north** / **south** of magnetic north.

Now look at and think about each of the *I can* statements.

Date: ________________

1 Label the diagram using the names of the Moon's different phases in the box.

last quarter	first quarter	waning crescent	waxing crescent
waning gibbous	waxing gibbous	full Moon	new Moon

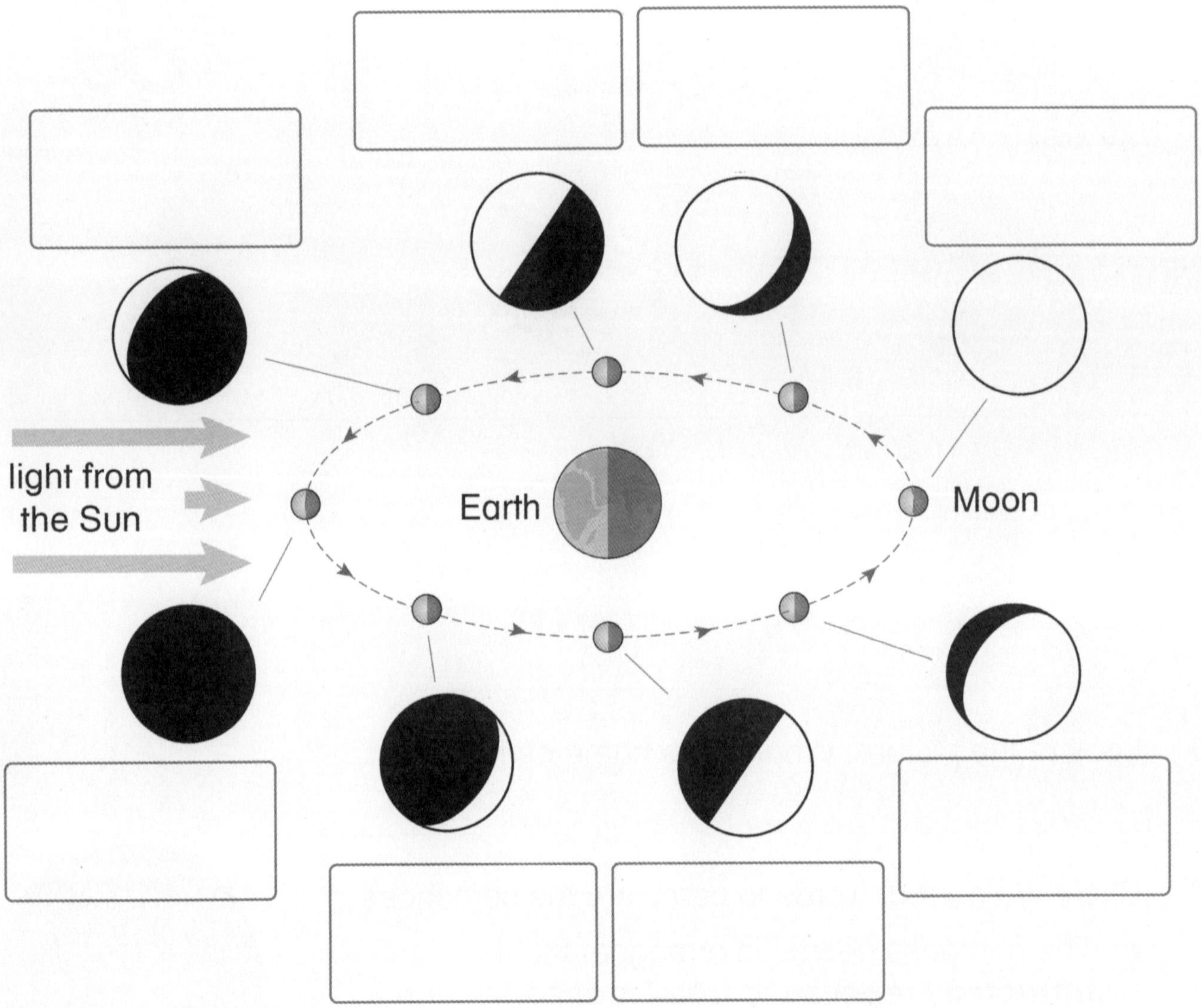

2 Draw a picture of a new Moon and a full Moon.

New Moon	Full Moon

Why do we see the Moon so differently at these two phases?

____

____

3 Use words from the box to complete the paragraph.

Sun lunar light sets spinning east
night orbits phases reflects west

The Moon is not a source of ______________. We see the Moon because it ______________ light from the Sun. Exactly half of the Moon is always lit up by the ______________, but because the Earth and Moon are both ______________ at different speeds, we see different amounts of the Moon lit up as it ______________ the Earth. This means that we see the Moon from a different angle every ______________ and so different shapes of the Moon are visible to us at different times of the ______________ month. These different shapes are known as the ______________ of the Moon. The Moon rises and ______________ as the Earth spins, which is why it looks like the Moon is moving across the sky at night. Just like the Sun, the Moon rises in the ______________ and sets in the ______________.

Date: ______________

Now look at and think about each of the *I can* statements.

1 Use the picture below to help you answer the questions.

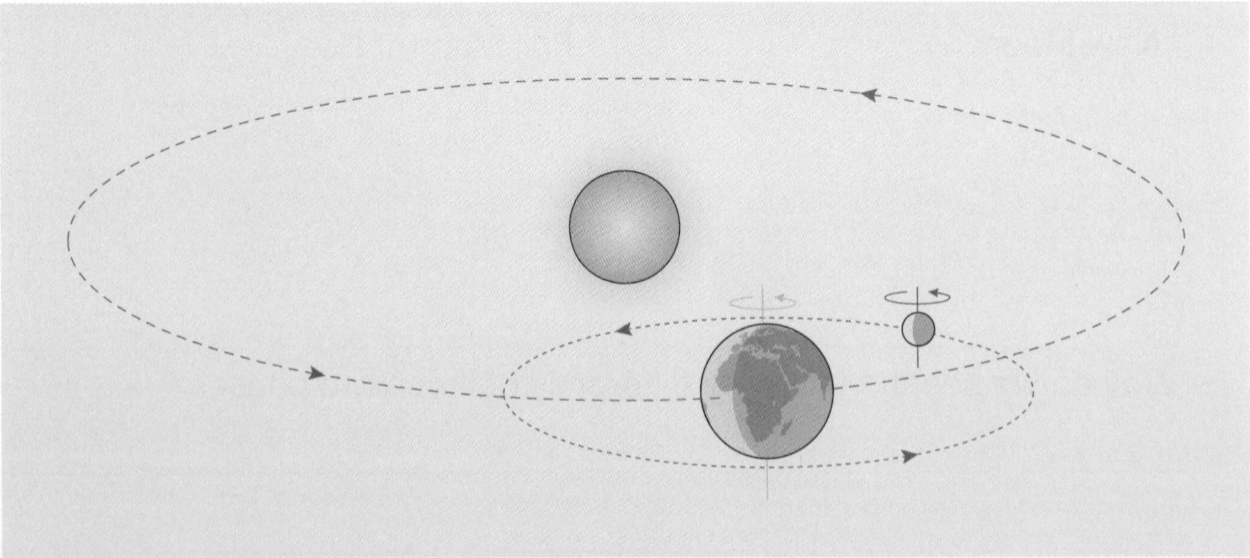

a In what direction does the Earth move?

____

b What is the shape of the Moon's orbit around Earth?

____

2 How long does it take for the Moon to complete a whole revolution from new moon to new moon?

____

3 How long does it take the Earth to rotate once?

____

4 How long does it take the Earth to complete one revolution around the Sun?

____

5 Look at the bar chart showing the hours of daylight in London in June and December. What do you observe from the bar chart?

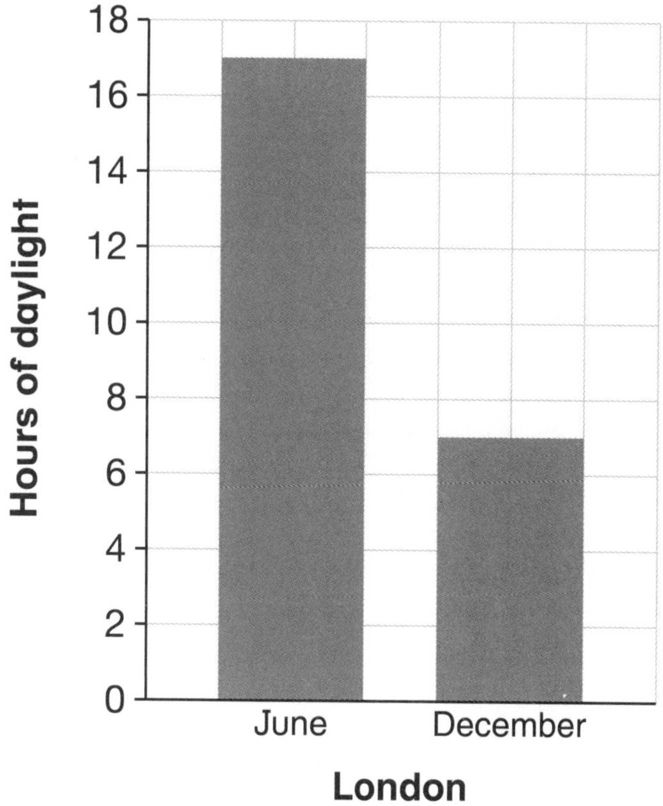

____

____

____

____

____

____

____

____

6 Draw lines to match the correct shapes of the Moon, Earth and Sun.

Moon, Earth and Sun
A Moon 
B Earth 
C Sun 

Shape
1 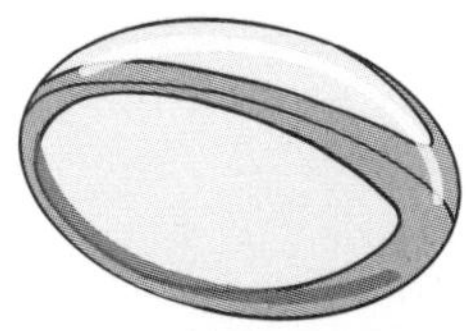
2 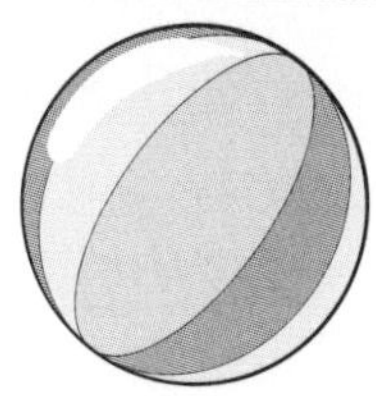
3 

Now look at and think about each of the *I can* statements.

Date: ______________________

1 What do we call scientists who study the Universe and all the objects in it?

____

2 What is the name of the system that people have been gathering information about for thousands of years?

____

3 What did the Italian scientist, Galileo, build to observe the objects in the skies?

____

4 What was the first object that was sent into space by the USSR (now Russia)?

____

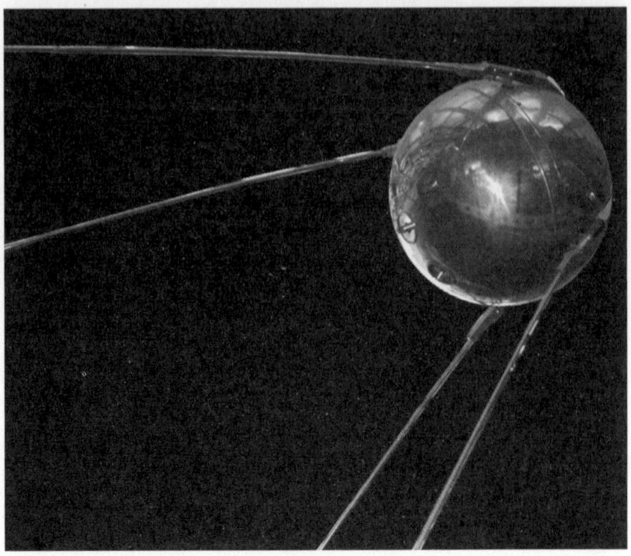

5 Was the first person in space the first person to walk on the Moon?

___

6 Who was the first person to walk on the Moon?

___

7 What do scientists do on the International Space Station (ISS)?

___

___

___

___

Date: ___

Now look at and think about each of the *I can* statements.

1 Look at the bar chart. Then answer the questions.

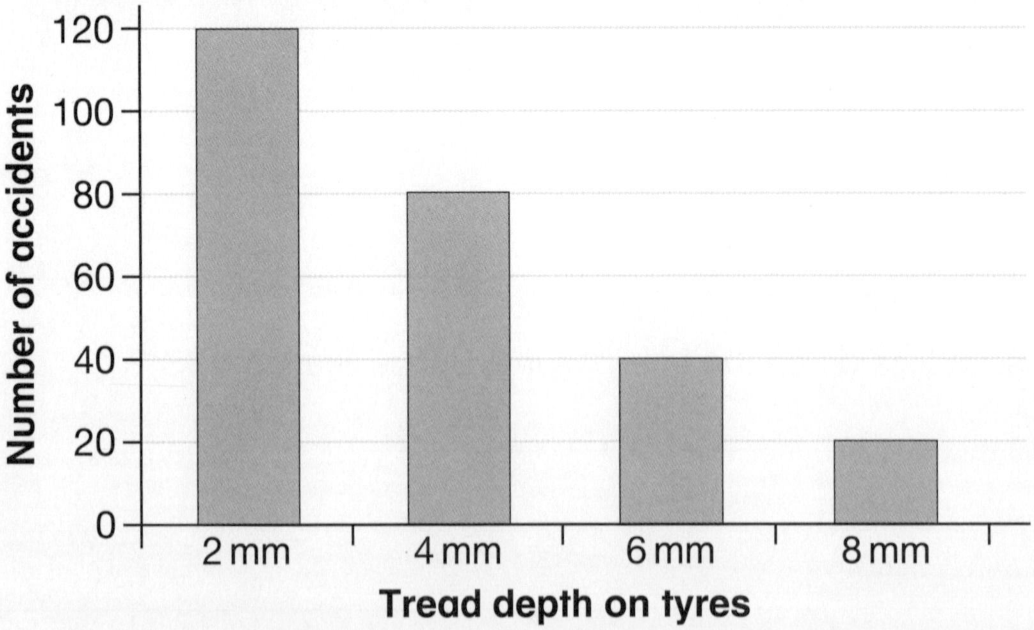

a What does the bar chart show when looking at the tread on car tyres and road accidents?

____

b Is there a pattern between the height of the bars and the amount of tread on the tyres?

____

2 Look at the picture of a car.

a Use a red arrow to show the direction the car is moving in.

b Use a blue arrow to show the force on the car.

c What do we call this force?

____

3 Why is the windscreen on a car transparent?

___

4 What would happen if the windscreen on a car was translucent?

___

5 Is the body of a car transparent, translucent or opaque?

___

6 Look at the picture. Draw the shadow of the car.

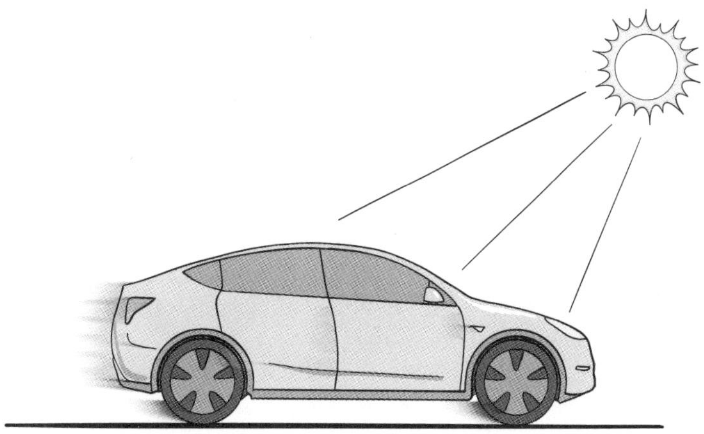

7 How are shadows formed?

___

8 What force causes snow to drop from a tree branch on to a car?

___

9 Circle the correct words to complete the sentences.

a The two ends of a magnet are known as the **south / east** pole and the **west / north** pole.

b Opposite poles of a magnet **attract / repel** each other, whereas the same poles **attract / repel** each other.

c Iron is a type of metal that is **non-magnetic / magnetic**. Plastic is an example of a material that is **non-magnetic / magnetic**.

d The needle in a compass is a magnet that is **attracted / repelled** to **true / magnetic** north.

10 Look at the picture. Write **new Moon** and **full Moon** in the correct places on the diagram.

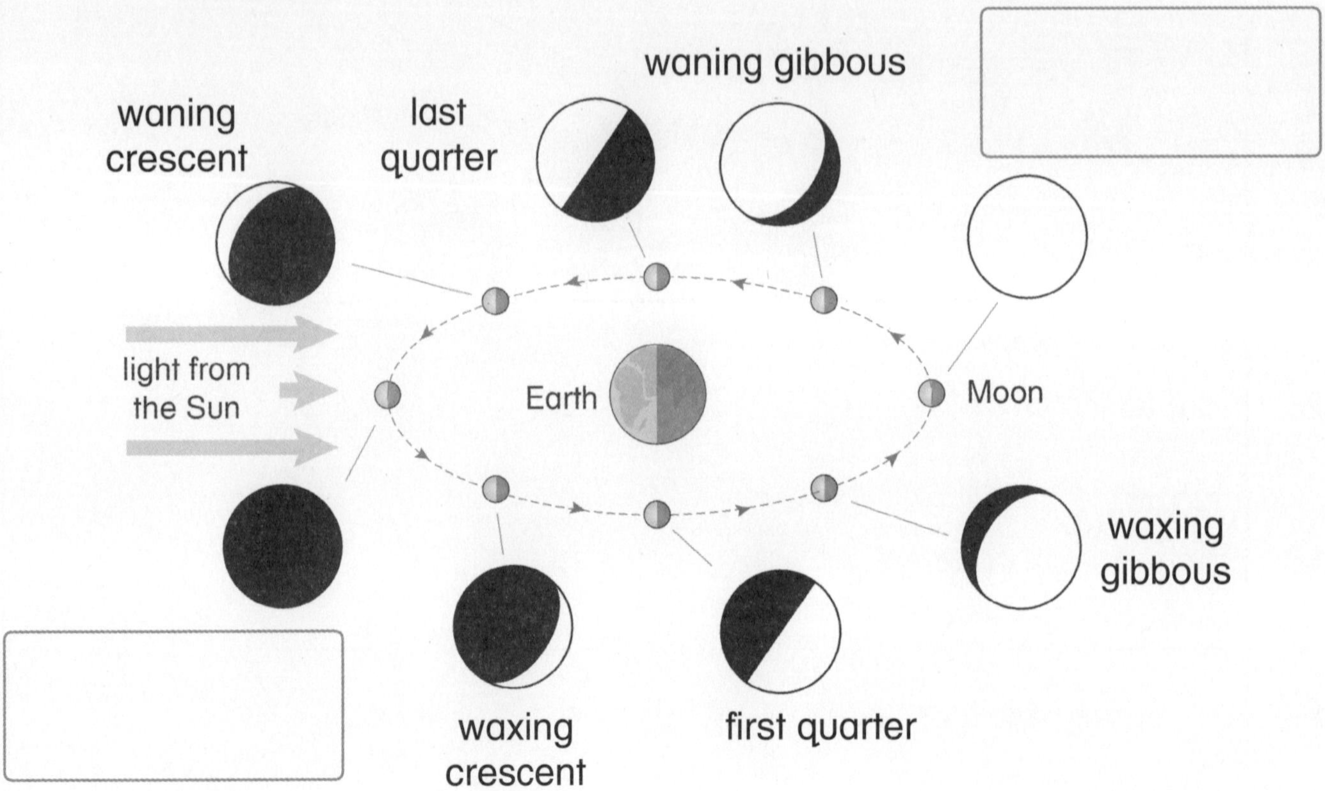

11 Where does the Moon get its light from?

____

12 Draw lines to match each time period to an occurrence.

Time period	Occurrence
29½ days	The Earth orbits the Sun once.
365¼ days	The Earth rotates anticlockwise once.
24 hours	The Moon orbits the Earth once.

13 Describe the shape of:

a the Moon – ____________________

b the Earth – ____________________

c the Sun – ____________________

14 Why does the Moon look different from Earth at different times of the lunar month?

____

15 From which direction does the Moon rise?

____

16 Where does the Moon set?

____

17 Where does the Sun rise?

____

18 Where does the Sun set?

____

Date: ______________________

Acknowledgements

Photo acknowledgements

The publishers gratefully acknowledge the permission granted to reproduce the copyright material in this book. Every effort has been made to trace copyright holders and to obtain their permission for the use of copyright material. The publishers will gladly receive any information enabling them to rectify any error or omission at the first opportunity.

P21t Ermess/Shutterstock; p21ct AlessandroZocc/Shutterstock; p21c Rainer Fuhrmann/Shutterstock; p21cb Jaroslav Mravcik/Shutterstock; p21b Welford Tiller/Shutterstock; p29 Glenn Young/Shutterstock; p32tl Yevgeniy11/Shutterstock; p32tr Wabeno/Shutterstock; p32bl voranat/Shutterstock; p32br cosma/Shutterstock; p33 Vilax/Shutterstock; p36b Mmzgombic/Shutterstock; p38tl FloridaStock/Shutterstock; p38tr LeonP/Shutterstock; p38cl Aastels/Shutterstock; p38cr Kwadrat/Shutterstock; p38bl Eric Isselee/Shutterstock; p38br Daniel Prudek/Shutterstock; p39t Eric Isselee/Shutterstock; p39ctt Daniel Prudek/Shutterstock; p39ct FloridaStock/Shutterstock; p39cb Aastels/Shutterstock; p39ccb Hayk_Shalunts/Shutterstock; p39b Kwadrat/Shutterstock; p40tl Ambient Ideas/Shutterstock; p40tr Cameramannz/Shutterstock; p40bl Sari Oneal/Shutterstock; p40br Breck P. Kent/Shutterstock; p43l zhukovvvlad/Shutterstock; p43r MarVil/Shutterstock; p53l The Toidi/Shutterstock; p53c Africa Studio/Shutterstock; p53r Vova Shevchuk; p56tl P Maxwell Photography/Shutterstock; p56tr Viktor1/Shutterstock; p59t Andrey Shtanko/Shutterstock; p59c Worradirek/Shutterstock; p59b Goory/Shutterstock; p60t Dorling Kindersley Ltd/Alamy Stock Photo; p60c New Africa/Shutterstock; p60b Marco Tulio/Shutterstock; p61tl Chones/Shutterstock; p61tc Ildi Papp/Shutterstock; p61tr Vitaly Korovin/Shutterstock; p61bl Jr images/Shutterstock; p61bc Bqmeng/Shutterstock; p61br Art_rich/Shutterstock; p64tl Outc/Shutterstock; p64tc Photobort/Shutterstock; p64tr Sakdinon Kadchiangsaen/Shutterstock; p64bl Yuris/Shutterstock; p64bc Koya979/Shutterstock; p64br Kaband/Shutterstock; p65l Valentyn Volkov/Shutterstock; p65c Africa Studio/Shutterstock; p65r Anton Starikov/Shutterstock; p71tl Iakov Filimonov/Alamy Stock Photo; p71tr Nail Bikbaev/Shutterstock; p71bl Angelika Smile/Shutterstock; p71br Ben Bryant/Shutterstock; p74t Dmitriy Kazitsyn/Shutterstock; p74ct Gino Santa Maria/Shutterstock; p74c Mikeledray/Shutterstock; p74cb Elena Elisseeva/Shutterstock; p74b Gresei/Shutterstock; p76bl Nodmitry/Shutterstock; p76bc Petr Malyshev/Shutterstock; p76br Avprophoto/Shutterstock; p77tl Somchai Som/Shutterstock; p77tc Phloem/Shutterstock; p77tr Magicoven/Shutterstock; p77ctl Stephen Rees/Shutterstock; p77ct Jocic/Shutterstock; p77ctr Jiang Dongmei/Shutterstock; p77cbl Oksana2010/Shutterstock; p77cb Picsfive/Shutterstock; p77cbr Lotus_studio/Shutterstock; p77bl Oleksiy Mark/Shutterstock; p77bc Vladimir Sazonov/Shutterstock; p77br Good Luck Photo/Shutterstock; p81tl Claudio Divizia/Shutterstock; p81br Maks Narodenko/Shutterstock; p82t Iryna1/Shutterstock; p82c Iryna/Shutterstock; p82b NASA Archive/Alamy Stock Photo; p83t Oleg_Kakovlev/Shutterstock; p83b Dima Zel/Shutterstock.